AF737139

Palgrave Studies in Maritime Politics and Security

Series Editor
Geoffrey F. Gresh, Springfield, VA, USA

The world's oceans cover over 70% of the planet's surface area. Global shipping carries at least 80% of the world's traded goods. Offshore oil and gas account for more than one-third of world energy production. With the maritime domain so important and influential to the world's history, politics, security, and the global political economy, this series endeavors to examine this essential and distinct saltwater perspective through an interdisciplinary lens, with a focus on understanding the ocean historically, politically, and from a security lens. Through a spectrum of engaging and unique topics, it will contribute to our understanding of the ocean, both historically and in a contemporary light, as source, avenue, and arena: a source of food and energy; an avenue for the flow of goods, people, and ideas; and an arena for struggle and warfare. The series will use an interdisciplinary approach—integrating diplomatic, environmental, geographic, and strategic perspectives—to explore the challenges presented by history and the contemporary maritime issues around the world.

More information about this series at
http://www.palgrave.com/gp/series/16675

Geir Hønneland

Blue Governance in the Arctic and Antarctic

Private Fisheries Certification and the Law of the Sea

palgrave
macmillan

Geir Hønneland
Nord University and Fridtjof Nansen Institute
Bodø and Lysaker, Norway

ISSN 2730-7972 ISSN 2730-7980 (electronic)
Palgrave Studies in Maritime Politics and Security
ISBN 978-3-030-72584-6 ISBN 978-3-030-72585-3 (eBook)
https://doi.org/10.1007/978-3-030-72585-3

© The Author(s), under exclusive license to Springer Nature Switzerland AG 2021
This work is subject to copyright. All rights are solely and exclusively licensed by the Publisher, whether the whole or part of the material is concerned, specifically the rights of translation, reprinting, reuse of illustrations, recitation, broadcasting, reproduction on microfilms or in any other physical way, and transmission or information storage and retrieval, electronic adaptation, computer software, or by similar or dissimilar methodology now known or hereafter developed.
The use of general descriptive names, registered names, trademarks, service marks, etc. in this publication does not imply, even in the absence of a specific statement, that such names are exempt from the relevant protective laws and regulations and therefore free for general use.
The publisher, the authors and the editors are safe to assume that the advice and information in this book are believed to be true and accurate at the date of publication. Neither the publisher nor the authors or the editors give a warranty, expressed or implied, with respect to the material contained herein or for any errors or omissions that may have been made. The publisher remains neutral with regard to jurisdictional claims in published maps and institutional affiliations.

Cover illustration: Nino Marcutti/Alamy Stock Photo

This Palgrave Macmillan imprint is published by the registered company Springer Nature Switzerland AG
The registered company address is: Gewerbestrasse 11, 6330 Cham, Switzerland

Preface

The research underlying this book partly emanates from the research project 'Spatial shifts of marine stocks and the resilience of polar resource management (STOCHSHIFT)', financed by the Research Council of Norway's Polar Research Programme (POLARPROG) and carried out at the University of Oslo and the Fridtjof Nansen Institute from 2016 to 2019. Thanks to project manager Olav Schram Stokke and fellow project participants Anne-Kristin Jørgensen and Andreas Østhagen for stimulating discussions on fisheries management in polar waters; to Professor Erik Molenaar at The Netherlands Institute for the Law of the Sea and the publisher's anonymous peer reviewer for detailed and constructive comments to the text; to Susan Høivik for assistance in keeping my language as clear and concise as possible; and to Professor Frode Mellemvik at the High North Center at Nord University for letting me use my part-time position there to prepare this manuscript. Parts of Chapters 3, 4 and 5 have been published as Geir Hønneland, 'Marine Stewardship Council (MSC) Certification of Arctic Fisheries: Processes and Outcomes', *Arctic Review on Law and Politics*, Vol. 11, 2020, pp. 133–156. Thanks to Professor Nigel Bankes at the University of Calgary and the two anonymous peer reviewers for useful comments to that manuscript.

For about a decade, I have been involved as an external expert in assessments according to the Marine Stewardship Council (MSC) Fisheries Standard. Since 2019, I have worked as fisheries assessment team leader for the global certification company Lloyd's Register. As this book will make clear, the MSC Fisheries Standard is a comprehensive scheme. Not least, it requires competence in marine sciences, in which I am not trained. I wish to extend my warmest thanks to my fellow MSC assessors and certification company staff who have continued to teach me what I need to know about fisheries certification—that is not done in one decade.

Oslo, Norway
January 2021

Geir Hønneland

Contents

List of Abbreviations

ACDR	Announcement Comment Draft Report (MSC)
ARK	Association of Responsible Krill Fishing Companies
ASC	Aquaculture Stewardship Council
ASI	Assurance Services International
ASOC	Antarctic and Southern Ocean Coalition
BAS	British Antarctic Survey
CAB	Conformity Assessment Body ('Certification Body')
CAP	Client Action Plan (MSC)
CCAMLR	Commission for the Conservation of Antarctic Marine Living Resources
CCRF	Code of Conduct for Responsible Fisheries (FAO)
COFI	Committee on Fisheries (FAO)
CPRDR	Client and Peer Review Draft Report (MSC)
DFFU	Deutsche Fischfang-Union
ETP	Endangered, Threatened and Protected (Species)
FAO	Food and Agriculture Organization of the United Nations
FCP	Fisheries Certification Process (Requirements) (MSC)
FSA	Fish Stocks Agreement (United Nations)
GSSI	Global Sustainable Seafood Initiative
HCR	Harvest Control Rule
ICES	International Council for the Exploration of the Sea
ICJ	International Court of Justice
IMR	Institute of Marine Research (Norway)

ISF	Iceland Sustainable Fisheries
ITLOS	International Tribunal for the Law of the Sea
IUU	Illegal, Unreported and Unregulated (Fishing)
MINSA	Mackerel Industry Northern Sustainability Alliance
MSC	Marine Stewardship Council
MSY	Maximum Sustainable Yield
NEAFC	North-East Atlantic Fisheries Commission
P1	Principle 1 (MSC) (Status of Target Stocks)
P2	Principle 2 (MSC) (Impacts on Wider Ecosystem)
P3	Principle 3 (MSC) (Effectiveness of Management System)
PCDR	Public Comment Draft Report (MSC)
PI	Performance Indicator (MSC)
PINRO	Nikolai M. Knipovich Polar Research Institute of Marine Fisheries and Oceanography (Russia)
PRI	Point Where Recruitment Would Be Impaired (MSC)
RFM	Responsible Fisheries Management (Standard)
RFMO	Regional Fisheries Management Organization
SARPC	Syndicat des Armements Réunionnais de Palangriers Congélateurs
SG	Scoring Guidepost (MSC)
SI	Scoring Issue (MSC)
TAC	Total Allowable Catch
UNCED	United Nations Conference on Environment and Development (1992)
UNCLOS	United Nations Convention on the Law of the Sea
UoA	Unit of Assessment (MSC)
UoC	Unit of Certification (MSC)
VCLT	Vienna Convention on the Law of Treaties
VMS	Vessel Monitoring System
WWF	World Wide Fund for Nature

Full references to international treaties are provided in Chapters 1 and 2.

List of Figures

List of Tables

CHAPTER 1

Introduction

Abstract Certification by private sustainability schemes has become a prerequisite for export—oriented fisheries around the world, and the most extensive and widespread scheme is that of the Marine Stewardship Council (MSC). Seafood exporters face not only lower prices for non-MSC-certified products: they are also effectively barred from some of the most lucrative global markets if their fish is not MSC-certified. Social science research on the MSC often focuses on the perceptions and effects of the MSC beyond fisheries management as such, addressing, inter alia, consumer willingness to pay for certified products, the legitimacy of the MSC Standard among stakeholders and the environmental, economic and social effects of MSC certification. By contrast, this book analyses the MSC Certification Scheme as a 'semi-legislative'/'regulatory' system and its links with international fisheries law, specifically the precautionary approach to fisheries management. It includes an investigation of the effects of MSC on fisher behaviour, national legislation and other forms of state practice.

Keywords Ecolabelling · Fisheries certification · Marine Stewardship Council

© The Author(s), under exclusive license to Springer Nature Switzerland AG 2021

G. Hønneland, *Blue Governance in the Arctic and Antarctic*, Palgrave Studies in Maritime Politics and Security,
https://doi.org/10.1007/978-3-030-72585-3_1

'Blue economy' is a buzzword in contemporary debates about ocean governance, at both academic and policy levels.[1] The concept seeks to couple traditional approaches to the management of marine resources with recent decades' increased focus on marine protection and, most importantly, quests for new synergies with marine-based businesses. In the fisheries sector, certification by private sustainability schemes has in recent years developed as a non-state alternative, or addition to, traditional management at national and international levels.

The golden standard of fisheries certification is accreditation by the Global Sustainable Seafood Initiative (GSSI),[2] and the first global scheme to achieve such recognition was the Fisheries Standard of the Marine Stewardship Council (MSC).[3] Since its establishment in 1997, the MSC has worked decisively to develop an increasingly rigorous standard for certification, and to get wholesale supply chains and retailers to commit to purchasing MSC-certified seafood only. As a consequence, seafood exporters face not only lower prices for non-MSC-certified products: they are effectively barred from the most lucrative global markets if their fish is not MSC-certified.

The MSC Certification Scheme—which consists of the MSC Fisheries Standard and the MSC Fisheries Certification Process (FCP) requirements (see Chapter 3)—applies only to a limited extent to fishing companies as such. The scheme primarily involves assessment of management systems, with requirements as to their outcome (e.g., status of target and bycatch stocks and other ecosystem components), management measures (e.g., harvest control rules and biological reference points) and availability of information (e.g., in the form of stock assessments).[4] Involving management authorities at national and international levels is necessary: it is the interaction between the fishing industry and management authorities that is meant to drive the sustainability of fisheries forward.

[1] See, e.g., Voyer et al. (2018) for an overview.

[2] See https://www.ourgssi.org/ (accessed 29 June 2020).

[3] See https://www.msc.org/ (accessed 29 June 2020). As per June 2020, 15% of the world's marine catch is MSC-certified.

[4] 'Outcome', 'management' and 'information' are MSC terminology, referring to three main categories of performance indicators against which fisheries are scored (see Chapter 3). Note that the word 'outcome' is used differently than in the analytical typology presented in the present chapter.

Questions to Be Asked

There is a burgeoning social science literature on private fisheries certification in general, and the MSC in particular. Many contributions focus on the perceptions and effects of the MSC beyond fisheries management as such, addressing, inter alia, consumer willingness to pay for certified products,[5] the legitimacy of the MSC Fisheries Standard among stakeholders[6] and the environmental, economic and social effects of MSC certification.[7] In the latter, the *impact* of certification (i.e. the achievement of final objectives, like improving the marine environment) is addressed, without first analysing the *output* (i.e. amendment of national laws, regulations and policies) or *outcome* (i.e. the influence on target, in this case fisher, behaviour).[8] This book does not address the wider impacts of MSC certification; nor does it primarily aim to communicate with the literature on private certification as such. Instead, it analyses the MSC Certification Scheme as a 'semi—legislative'/'regulatory' system and its links with international fisheries law, specifically the precautionary approach to fisheries management. It includes an investigation of the effects of MSC on fisher behaviour, national legislation and other forms of state practice.

The MSC Fisheries Standard contains numerous references to international fisheries law, such as the 1982 United Nations (UN) Convention on the Law of the Sea (UNCLOS),[9] the 1995 UN Fish Stocks Agreement

[5] See, e.g., Lim et al. (2018).

[6] See, e.g., Gulbrandsen and Auld (2016).

[7] See, e.g., Arton et al. (2020). This is just a small selection of topics covered in the literature about MSC. A search in the ISI Web of Science (www.webofknowledge.com) (accessed 21 April 2020) yielded 161 articles, 131 of them published since 2012. Many of these are empirical studies of MSC assessments in specific geographic regions. I have not found other overview articles of MSC assessments in the Antarctic or the Arctic, but some Russian Barents Sea fisheries are analysed in articles about MSC assessments in Russia (Gulbrandsen and Hønneland 2014; Pristupa et al. 2016; Lajus et al. 2018). It should be noted that it is not the purpose of this book to engage primarily with the literature on private certification more widely, but more with the literature on the implementation and application of the precautionary approach in international fisheries management, and specifically how this can be achieved through private certification.

[8] See Chapter 2. The typology *output/outcome/impact* is taken from Andresen et al. (2012, pp. 7–8).

[9] United Nations Convention on the Law of the Sea, Montego Bay, 10 December 1982, U.N.T.S. Vol. 1833, p. 3.

(FSA)[10] and the 1995 FAO Code of Conduct for Responsible Fisheries (CCRF),[11] in particular regarding the precautionary approach to fisheries management. The overarching research question of this book is: *to what extent and in what way do private standards like the MSC Fisheries Certification Scheme contribute to the implementation of the precautionary approach to fisheries management?* As explained in Chapter 2, the term 'implementation' is used in the literature in narrow and broad forms. The narrow version sees implementation as referring to the transformation or incorporation of international obligations (relating to legally binding rules) and commitments (relating to non-legally binding rules) into national law only (*output*, in the terminology above). The broader version comprises changes in other forms of state practice beyond the introduction and/or amendment of national legislation, including the operationalization[12] and application of general concepts in concrete circumstances (also *output*), as well as target behaviour aimed at effectuating the requirements of the international instrument/national legislation (*outcome*). Further, conceptions differ as to who can be actors in implementing activities. In the narrow version, only states may implement international agreements, as national legislation is their remit. In the broad version, other actors can contribute to the implementation process, typically non-governmental organizations (NGOs) and international organizations, and implementation is understood as all activities deliberately and explicitly aimed at bringing the behaviour of target groups into line with the requirements of international agreements (and national legislation insofar as it reflects international commitments). It is the broader version of implementation that is used in this book.

The overarching research question can be split into two more specific research questions:

1. To what extent and in what form has the precautionary approach been incorporated in the MSC Certification Scheme?

[10] Agreement for the Implementation of the Provisions of the United Nations Convention on the Law of the Sea of 10 December 1982 relating to the Conservation and Management of Straddling Fish Stocks and Highly Migratory Fish Stocks, New York, 4 August 1995, U.N.T.S. Vol. 2167, p. 3.

[11] Code of Conduct for Responsible Fisheries, FAO, Rome, 31 October 1995.

[12] By 'operationalization' is meant the concretization of vague or generic concepts into more specific and more easily observable (and ideally measurable) units.

2. To what extent and in what way has MSC certification affected target (fisher) behaviour and state practice in the application of the precautionary approach?

The empirical study of the second research question focuses on four clusters of fisheries in polar waters: one in the Antarctic and three in the Arctic. In the Southern Ocean, the commercially most important species are toothfish, which is a slow-growing deep-water top predator, and krill, a pelagic (i.e. moving in the water column) low-trophic species with a central role in the food web. In the Barents Sea, the main fishery is for demersal stocks (i.e. stocks living on the relatively shallow continental shelf), of which cod and haddock are the economically most important. In the Norwegian Sea, pelagic species dominate, i.e. species living in the water column and moving in shoals, like mackerel, herring and blue whiting—commercially important, but short-lived and fast-moving. In this book, we focus on mackerel fisheries. The fourth cluster is inshore fisheries for three stocks of one species, Arctic lumpfish, in Greenland, Iceland and Norway. Lumpfish is a semi-pelagic species caught primarily for its roe.

These four clusters of fisheries are managed at the multilateral, bilateral and national levels, respectively. Fisheries in the Southern Ocean are managed multilaterally by the Commission for the Conservation of Marine Living Resources (CCAMLR), established in 1980 and innovative in its early application of the precautionary and ecosystem-based approaches to fisheries management. In the Barents Sea, the most important stocks are managed by the Joint Norwegian–Russian (formerly Soviet) Fisheries Commission. Set up in 1975 and involving only two states, the regime is tightly integrated, mature and generally considered to be effective. By contrast, the pelagic stocks in the Norwegian Sea are subject to looser and more recent management arrangements, the so—called coastal states agreements between the EU, Norway, the Faroe Islands, and in recent years to some extent also Iceland and Greenland (and Russia, as a non—coastal state recognized as having historical rights in the area). The North—East Atlantic Fisheries Commission (NEAFC) also has a role in the management of the stocks in areas beyond national jurisdiction in the Norwegian Sea ('the Banana hole').[13]

[13] See the websites of CCAMLR (www.ccamlr.org), the Joint Norwegian–Russian Fisheries Commission (www.jointfish.org) and NEAFC (www.neafc.org) (all accessed 24

There are practical and analytical limitations to this study. My original intention to cover other provisions in international fisheries law beyond the precautionary approach (e.g. the ecosystem approach and compliance-related rules) and a wider selection of polar fisheries was abandoned, likewise, my plans to supplement the documentary review with personal interviews with key stakeholders. These limits to the depth and breadth of the empirical investigation affect the validity (the certainty that my empirical conclusions are correct) and the generalizability of this study. I add an analytical reservation concerning causation: while it may be possible to point empirically to a chronological link between MSC certification, fisher behaviour and state practice (and sources may claim there *is* a causal link), it is beyond the ambition of this book to test alternative explanations to the observed behaviour.

Sources and Methods

The study combines legal analysis with empirical investigation of the MSC Fisheries Certification Scheme programme documents and assessment reports. A traditional doctrinal method is applied in defining the content of the precautionary approach in international fisheries law, in accordance with the recognized sources of international law pursuant to Article 38 of the Statute of the International Court of Justice (ICJ),[14] and the treaty interpretation rules in Articles 31–32 of the Vienna Convention on the Law of Treaties (VCLT).[15] Here I analyse binding treaties (ICJ Statute, Article 38(1)(a)), in particular, the FSA; and, as a subsidiary means of interpretation, the work of publicists (ICJ Statute, Article 38(1)(d)). I also review, again as a potential subsidiary means, relevant case law from the ICJ, the International Tribunal for the Law of the Sea (ITLOS) and arbitral tribunals established under Annex VII of UNCLOS (ICJ Statute, Art. 38(1)(d)). Further, I analyse soft law instruments such as the CCRF, which do not carry weight as sources of law but might provide useful

August 2020). An updated discussion of fisheries management in the Southern Ocean and the Northeast Atlantic is found in Haward (2020).

[14] Statute of the International Court of Justice (Charter of the United Nations and Statute of the International Court of Justice), San Francisco, 24 October 1945.

[15] Vienna Convention on the Law of Treaties (with Annex) Concluded at Vienna on 23 May 1969, U.N.T.S. Vol. 1155, p. 331.

context for the interpretation of legislation, enforcement action and judgments of national courts.[16] Importantly, I examine state practice, which is a term commonly used in reference to 'general practice' as one of the criteria for a rule to qualify as a rule of customary international law (ICJ Statute, Art. 38(1)(b)).[17] For that to be the case, the practice must be widespread and representative ('general'), consistent and accompanied by a recognition that the practice reflects a rule of law[18] (see also Chapter 2).

It is the *content* of the precautionary approach which is at issue in this book, not its potential status as reflecting customary international law (although I refer briefly to the views of international courts and publicists on this question in footnotes in Chapter 2). In any event, the empirical coverage of the investigation is far too narrow to enable drawing any conclusions on the legal status of the approach under international law. Hence, my focus is on mapping fisher behaviour and state practice empirically, which may then subsequently (i.e. beyond this book) contribute to the discussion of the legal status of the precautionary approach under international law together with a far wider selection of instances of state practice from other empirical settings.

An empirical investigation of the MSC Fisheries Standard and assessment reports, which underlies the more political science—oriented study of MSC implementation of the precautionary approach, draws on traditional qualitative methods in social sciences: a 'factual' variant of textual analysis (as opposed to, e.g., a discourse analytical approach).[19] The study of implementation of international environmental agreements has a firm tradition of combining international law and political science (specifically, international relations) perspectives.[20] This book follows in that tradition.

A rather extensive empirical investigation has been carried out, the results of which are summarized in the tables in Chapter 4 and presented

[16] Non-binding treaties cannot count as relevant context for interpretation in accordance with the VCLT, Art. 31(1), since Art. 31(2)(a)(b) and Art. 31(3)(a) relate context to other *binding* instruments between (all) the parties, and none of the other provisions apply either.

[17] Wood and Sender (2017, p. 1, Art. 2).

[18] Ibid., p. 1, Art. 2.

[19] See, e.g., Silverman (2017).

[20] Two classic works on the implementation of international environmental agreements which combine international law and international relations perspectives are Weiss and Jacobson (1998) and Victor et al. (1998).

in more detail in that chapter and Chapter 5. As the information on fisheries' improvement conditions, stakeholder comments, objections and efforts to meet conditions following certification (see Chapter 3) is not generally available in aggregate form, I have examined all assessment and re-assessment reports, as well as annual surveillance audit reports (between 500 and 1000 pages per five-year certification period for each fishery), for all the 20 fisheries[21] in the study (relating to seven different stocks of five different species). The essential information is tabulated, and reference to all assessment reports is provided in Appendix 1.

I am myself a certified MSC auditor and have been involved in fishery assessments as an expert on MSC Principle 3 (the management system) and assessment team leader.[22] This practical experience and training in applying the MSC Standard have helped me navigate the voluminous body of documentation available on MSC assessments. The empirical investigation of this book is based solely on publicly available documents—hence, I see no conflict of interest or other ethical problems.[23]

Outline of the Book

The precautionary approach and the concept of implementation are presented in Chapter 2. Chapters 3 and 4 are the main empirical chapters of the book. Chapter 3 gives an overview of the MSC Certification Scheme—the procedural and substantive rules that govern MSC assessments—and discusses to what extent and in what form the precautionary approach is incorporated there. Chapter 4 presents the results of the MSC assessment of the selected polar fisheries, including information about the assessment process, such as stakeholder submissions and objections.

[21] A 'fishery' is the standard unit assessed under the MSC scheme; see Chapter 3.

[22] This includes the Aker Biomarine krill fishery, the Russian Barents Sea fisheries, the Faroese and the Mackerel Industry Northern Sustainability Alliance (MINSA) mackerel fisheries, and the Norwegian and Icelandic Arctic lumpfish fisheries (second attempt) discussed in this book.

[23] At a more general level, one might ask whether my involvement with MSC certification affects my interpretation of the data. If so, it is not obvious whether that would make me less critically disposed, or more. In line with general standards for qualitative research, the best I can do is to make my situatedness explicit.

Chapter 5, the main analytical chapter, starts with an empirical discussion of the extent to which MSC certification has affected the behaviour of fishers and the management practices of the states involved. Then it reverts to the theoretical discussion initiated in Chapter 2, analysing the role of private certification as an implementation mechanism in international fisheries governance, and what forms of the precautionary approach are reflected in the MSC Standard and in MSC application of it. The main conclusions are summed up in Chapter 6.

References

Andresen, Steinar, Elin Lerum Boasson and Geir Hønneland, 'An International Environmental Policy Takes Shape', in Steinar Andresen, Elin Lerum Boasson and Geir Hønneland (eds), *International Environmental Agreements: An Introduction*, Abingdon: Routledge, 2012.

Arton, Ashleigh, Anthony Leiman, Gillian Petrokofsky, Hilde Toonen and Catherine S. Longo, 'What Do We Know About the Impacts of the Marine Stewardship Council Seafood Ecolabelling Program? A Systematic Map', 9 *Environmental Evidence* 2020, pp. 1–20.

Gulbrandsen, Lars H. and Graeme Auld, 'Contested Accountability Logics in Evolving Nonstate Certification for Fisheries Sustainability', 16 *Global Environmental Politics* 2016, pp. 42–60.

Gulbrandsen, Lars H. and Geir Hønneland, 'Fisheries Certification in Russia: The Emergence of Non-State Authority in a Postcommunist Economy', 45 *Ocean Development and International Law* 2014, pp. 341–359.

Haward, Marcus, *Governing Oceans in a Time of Change: Fishing for the Future?*, Cheltenham: Edward Elgar, 2020.

Hewison, Grant J., 'The Precautionary Approach to Fisheries Management: An Environmental Perspective', 11 *International Journal of Marine & Coastal Law* 1996, pp. 301–332.

Lajus, Dmitry, Daria Stogova and E. Carina H. Keskitalo, 'The Implementation of Marine Stewardship Council (MSC) Certification in Russia: Achievements and Considerations', 90 *Marine Policy* 2018, pp. 105–114.

Lim, Kar H., Wuyang Hu and Rodolfo M. Nayga Jr., 'Is Marine Stewardship Council's Ecolabel a Rising Tide for All? Consumers' Willingness to Pay for Origin-Differentiated Ecolabel Canned Tuna', 96 *Marine Policy* 2018, pp. 18–26.

Pristupa, Alexey O., Machiel Lamers and Bas Amelung, 'Private Informational Governance in Post-Soviet Waters: Implications of the Marine Stewardship Council Certification in the Russian Barents Sea Region', 182 *Fisheries Research* 2016, pp. 128–135.

Silverman, David, *Doing Qualitative Research*, Thousand Oaks, CA: Sage, 2017.
Victor, David G., Kal Raustiala and Eugene B. Skolnikoff (eds), *The Implementation and Effectiveness of International Environmental Commitments: Theory and Practice*, Cambridge, MA: MIT Press, 1998.
Voyer, Michelle, Genevieve Quirk, Alistar Mcllgorm *et al.*, 'Shades of Blue: What Do Competing Interpretations of the Blue Economy Mean for Oceans Governance?', 20 *Journal of Environmental Policy & Planning 2018*, pp. 595–616.
Weiss, Edith Brown and Harold K. Jacobson (eds), *Engaging Countries: Strengthening Compliance with International Environmental Accords*, Cambridge, MA: MIT Press, 1998.
Wood, Sir Michael and Omri Sender, 'State Practice', *Max Planck Encyclopedias of International Law* (Module: Max Planck Encyclopedia of Public International Law), last updated January 2017.

CHAPTER 2

The Precautionary Approach, Implementation and State Practice

Abstract The term 'implementation' is used in the literature in narrow and broad forms. In brief, the narrow version sees implementation as referring to the transformation or incorporation of international obligations (with legally binding rules) and commitments (with non-legally binding rules) into national law only, whereas the broader version also includes changes in other forms of state practice (beyond the introduction and/or amendment of national legislation, including the application of general concepts in concrete circumstances) and target behaviour aimed at effectuating the requirements of the international instrument/national legislation. Further, conceptions differ as to who can be actors in implementing activities. In the narrow version, only states may implement international agreements, as national legislation is their remit. In the broad version, other actors can contribute to the implementation process, typically non-governmental organizations (NGOs) and international organizations, and implementation is understood as all activities deliberately and explicitly aimed at bringing the behaviour of target groups into line with the requirements of international agreements, and national legislation insofar as it reflects international commitments.

Keywords Ecolabelling · Marine Stewardship Council · Precautionary approach · Implementation · State practice

© The Author(s), under exclusive license to Springer Nature Switzerland AG 2021

G. Hønneland, *Blue Governance in the Arctic and Antarctic*, Palgrave Studies in Maritime Politics and Security, https://doi.org/10.1007/978-3-030-72585-3_2

The first section of this chapter presents the emergence of the precautionary approach in international law, specifically in international environmental law, including how different views on the approach have been taken by publicists.[1] In the second section, the adoption and operationalization of the approach in international fisheries law is presented, with emphasis on the FSA and the CCRF. In the third section, I turn to the analytical approaches to understanding the interconnectedness of implementation, target behaviour and state practice. In particular, a conception of implementation is fleshed out that is broader than plain incorporation or transformation of a state's international obligations into national legislation.

The Precautionary Approach in International Law

Expositions of the precautionary principle tend to start with reference to the *Vorsorgeprinzip* or *Vorsorgegebot*, introduced in West German environmental law in the 1970s.[2] The principle found its way into international law in connection with German proposals to the International North Sea Ministerial Conferences.[3] It was first explicitly formulated in the 1987 Declaration of the Second International Conference on the Protection of the North Sea (the London Declaration)[4] and was adopted as a premise for future work in the 1990 Declaration of the Third International Conference on the Protection of the North Sea (the Hague Declaration).[5] Its geographical scope was extended to the wider Northeast Atlantic through the work of the Oslo and Paris Commissions, which resulted in the incorporation of the precautionary principle in the 1992

[1] The legal status of the precautionary approach will not be discussed in any detail here, but some brief comments are given in footnotes on how publicists and international courts and tribunals have assessed the legal status of the approach.

[2] See, e.g., Freestone and Hey (1996, p. 4), Ebben (2011, p. 115) and Sands and Peel (2018, p. 230).

[3] Freestone and Hey (1996, p. 5). This paragraph builds on Hønneland (2004, pp. 8–9).

[4] Ministerial Declaration on the Protection of the North Sea, 14(4) *Environmental Conservation* 1987, pp. 357–361. https://doi.org/10.1017/S0376892900016878.

[5] Ministerial Declaration of the Third International Conference on the Protection of the North Sea, The Hague, 8 March 1990, available at https://www.eea.europa.eu/policy-documents/ministerial-declaration-of-the-third (accessed 23 August 2020).

Convention for the Protection of the Marine Environment of the North-east Atlantic (OSPAR Convention).[6] Subsequently, the precautionary approach[7] was adopted in various international environmental agreements before becoming a pillar of the Rio Declaration[8] at the United Nations Conference on Environment and Development (UNCED) in 1992. The precautionary approach was also included in the two flagship treaties adopted at UNCED: the Convention on Biological Biodiversity[9] and the United Nations Framework Convention on Climate Change.[10]

The precautionary approach is related to the threshold for action (i), seen in relation to the level of scientific certainty (ii), in a situation of a threat of serious or irreversible environmental damage (iii). The relationship between these three elements, however, is not given. It has been discussed in the academic literature, as has the principle's status in international law. Several scholars emphasize the 'levels of strictness' to the principle. Sands and Peel note that, at the most general level, the precautionary principle implies that states 'agree to act carefully and with foresight when taking decisions that concern activities which may have an adverse impact on the environment'.[11] A more focused interpretation holds that specific activities or substances are regulated, perhaps banned, even if there is no conclusive evidence that harm may be caused to the environment.[12] An even stricter interpretation shifts the traditional burden of proof: while previously the burden of proof lay with those who

[6] Convention for the Protection of the Marine Environment of the North-East Atlantic (OSPAR Convention), Paris, 29 September 1992, U.N.T.S. Vol. 2354, p. 67.

[7] Whereas the precautionary *principle* is used in the OSPAR Convention and other international environmental agreements, the precautionary *approach* is used in the Rio Declaration and, as we shall see in the next section, international fisheries agreements. For the sake of consistency, I will use *approach* in the following, except when referring to an author who uses *principle*.

[8] Rio Declaration on Environment and Development, *Report of the United Nations Conference on Environment and Development*, Annex 1, Rio de Janeiro, 3–14 June 1992.

[9] Convention on Biological Diversity, Rio de Janeiro, 5 June 1992, U.N.T.S. Vol. 1760, p. 79. The Convention does not use the word 'precautionary', but the text reflects generally accepted definitions of the principle; cf. Preamble.

[10] United Nations Framework Convention on Climate Change, New York, 9 May 1992, U.N.T.S. Vol. 1771, p. 107. The Convention was not signed at UNCED, but was opened for signature at UNCED on 3 June 1992.

[11] Sands and Peel (2018, p. 234).

[12] Ibid., p. 234.

claimed an activity might cause harm to the environment, now those who conduct such activities must prove that no harm will be caused.

Similarly, Gillespie refers to 'weak' and 'strong' versions of the precautionary principle.[13] The former is accompanied by 'an optimistic assumption' about the renewability of environmental resources; the role of precaution is to function as 'a spur to innovation and managerial adaptation'.[14] This version 'pushes the burden of proof out a bit, into the anticipation of possible harm or expected proof of harm'.[15] The strong version, on the other hand, is supported by 'a pessimistic world view, whereby precaution is necessary because the ecology is not believed to be robust, and requires much greater intervention than has been achieved in the past'.[16]

Ebben provides an overview of different positions taken by publicists as regards the content and usefulness of the principle.[17] A common thread among the selected authors in his study is the categorization of the principle into 'strong' and 'weak' variants; they differ in their views on its status in international law, and its attractiveness. Similar to Sands and Peel above, Bodansky speaks of three variants of the precautionary principle: (i) the basic version, which functions as a reason for not postponing protective measures on grounds of scientific uncertainty; (ii) a stronger version, which gives states licence to take action; and (iii) the strongest version, which emplaces on states the duty to act, potentially changing the burden of proof.[18] Bodansky identifies three triggers of preventive action: the severity of the potential harm, which he deems the most important one; the likelihood of that harm; and the source of the harm.[19] However,

[13] Gillespie (2007, pp. 71–72).

[14] Ibid., p. 71.

[15] Ibid., p. 71.

[16] Ibid., p. 71. The weak and strong versions differ as to 'the nature of the threat, the triggering point for the principle, and the reversal of the burden of proof' (ibid., p. 71). As application of the principle will rarely have all three of these on one side, attempts to ascertain the legal status of the principle should veer towards the weak version, according to Gillespie (ibid., p. 74).

[17] Ebben (2011, pp. 117–124).

[18] Bodansky (2004, pp. 380–386), referred to in Ebben (2011, pp. 120–121).

[19] Bodansky (2004, p. 387), referred to in Ebben (2011, p. 121).

he sees the precautionary principle as lacking a clear definition and being unclear in its usefulness.[20]

Hence, there is no clear definition of the precautionary approach in international environmental law. Publicists tend to emphasize that there are different versions of the approach, varying in the degree of strictness. Some argue that the broader interpretations of the approach can be accorded customary status, at least within certain functional fields and/or geographical regions; others dismiss it.[21] International courts and tribunals have been reluctant to accept the precautionary principle as reflecting international customary law, but acceptance seems to be increasing.[22]

[20] Bodansky (2004, p. 391), referred to in Ebben (2011, p. 121).

[21] As an example, Sands and Peel (2018, p. 239) note that there is sufficient evidence in state practice to argue that the precautionary principle reflects a rule of customary international law, at least within the context of the European Union. More widely, the principle shall, in their opinion, at least be applied in the interpretation of international instruments (ibid., p. 240).

[22] The precautionary principle was first invoked before the ICJ in New Zealand's 1995 request concerning French nuclear testing, but the Court did not address the principle (Request for an Examination of the Situation in Accordance with Paragraph 63 of the Court S Judgment of 20 December 1974 in the Nuclear Tests (New Zealand v. France) Case, ICJ Reports 1995, p. 288). In the 1997 Gabčíkovo-Nagymaros judgement, the ICJ referred to the precautionary principle as among new norms and prescriptions in international environmental law (Gabčíkovo-Nagymaros Project (Hungary/Slovakia), Judgment, ICJ Reports 1997, p. 7); while in the 1999 Southern Bluefin Tuna case, the Annex VII arbitral tribunal prescribed a precautionary approach in all but name (Southern Bluefin Tuna (Australia and New Zealand v. Japan), 39 ILM 1359 (2000), 4 August 2000). The ICJ took one step further in the 2010 Pulp Mills case, noting that while a precautionary approach may be relevant in the interpretation and application of a treaty, that does not mean that it operates as a reversal of the burden on proof (Pulp Mills on the River Uruguay (Argentina v. Uruguay), Judgment, ICJ Reports 2010, p. 14). The 2011 Seabed Chamber Advisory Opinion, in turn, noted that the precautionary approach was on the way to achieving customary status (Responsibilities and Obligations of States with Respect to Activities in the Area, Advisory Opinion, 1 February 2011, ITLOS Reports 2011, p. 10).

The Precautionary Approach to Fisheries Management

At the time of the UNCED, there was little support within the fishing domain for applying the emerging precautionary approach, or precautionary principle, as it was often called, to fisheries management.[23] As noted by former Director of the FAO Fisheries and Aquaculture Management Division Serge M. Garcia in 1994, the principle could offer an opportunity to improve fisheries management and ensure sustainable fisheries developments, while '[i]ts careless generalization to fisheries could, however, lead to economic and social chaos in the fishing industry'.[24] Post-UNCED consultations within the FAO led to an agreement that the precautionary principle should not be applied to fisheries management, as fishing is an activity fundamentally different from more damaging practices such as toxic waste dumping, for which the principle had originally been adopted.[25] The FAO eventually came to advocate the precautionary *approach* for fisheries management as an alternative to the precautionary *principle*. The approach was considered a less stringent variant of the principle, entailing more flexible application of precautionary measures.[26]

The FSA was adopted in 1995.[27] Article 6, titled 'Application of the precautionary approach', contains seven paragraphs. According to Article 6(1), states 'shall apply the precautionary approach widely […] in order to protect the living marine resources and preserve the marine environment'. Article 6(2) declares: 'States shall be more cautious when information is uncertain, unreliable or inadequate. The absence of adequate scientific information shall not be used as a reason for postponing or failing to take conservation and management measures'. Article 6(3) lists steps that states shall take when implementing the precautionary approach:

[23] This paragraph builds on Hønneland (2004, pp. 10–11).

[24] Garcia (1994, p. 100).

[25] Ibid., p. 100.

[26] For an overview of these developments, see Kaye (2000, pp. 167–168, 187–197, 205–232).

[27] While the FSA in general applies to the conservation and management of straddling and highly migratory fish stocks beyond areas of national jurisdiction, Art. 6 (as well as the General principles in Art. 5 and the provisions on compatibility of management measures in Art. 7) applies equally to the conservation and management of such stocks within areas of national jurisdiction (Art. 3(1), 3(2)).

(i) obtain and share the best scientific information available and implement improved techniques for dealing with risk and uncertainty (Article 6(3)(a)); (ii) determine, on the basis of the best scientific information available, stock-specific reference points and the actions to be taken if these are exceeded (Article 6(3)(b)); (iii) take into account a range of uncertainties related to, inter alia, the status and dynamics of the fish stocks and their relation to established reference points, the impact on non-target and associated or dependent species, as well as oceanic, environmental and socio-economic conditions (Article 6(3)(c)); and (iv) develop data collection and research programmes to assess the impact of fishing on non-target and associated or dependent species and their environment and adopt plans to ensure the conservation of such species and the protection of habitats of special concern (Article 6(3)(d)).

FSA Annex II defines a precautionary reference point as 'an estimated value derived through an agreed scientific procedure, which corresponds to the state of the resource and of the fishery, and which can be used as a guide for fisheries management' (Para. 1). Two types of precautionary reference points shall be used: conservation, or limit, reference points; and management, or target, reference points (Para. 2). The former sets boundaries intended to contain harvesting within safe biological limits within which stocks can produce maximum sustainable yield (MSY) (ibid.). The latter reference points are intended to meet management objectives. Management strategies shall seek to maintain or restore stocks at levels consistent with agreed target reference points, with measures that can be implemented when reference points are approached (Para. 4). Fisheries management systems should aim to ensure that target reference points are, on average, not exceeded (Para. 5). The fishing mortality rate—in practice: the catch rate—that generates the MSY is to be regarded as a minimum standard for limit reference points (Para. 7).

The CCRF, adopted at the 28th Session of the FAO Conference in October/November 1995, is sometimes perceived as an implementing tool for the FSA, which was concluded a few months earlier. However, the initiative to the CCRF was taken by the FAO Committee on Fisheries (COFI) in March 1991—nearly two years before negotiations on the FSA were initiated in December 1992.[28] The CCRF is more correctly perceived as one of three components in the bundle of important legally

[28] See Annex I to the CCRF for an overview of its history.

binding and non-legally binding instruments on international fisheries management adopted around the mid-1990s, the other two being the FSA and the 1993 FAO Compliance Agreement.[29]

The CCRF consists of 12 extensive articles, of which Article 6 on general provisions and Article 7 on fisheries management are relevant here.[30] By way of introduction, Article 6.5 reiterates Articles 6(1), 6(2) and 6(3)(c) of the FSA. Article 7.5 on the precautionary approach contains five paragraphs, of which 7.5.1 and 7.5.2 also repeat, almost verbatim, the same provisions as the FSA's Article 6.5; Articles 7.5.4 and 7.5.5 reiterate FSA Articles 6(6) and 6(7), respectively. Article 7.5.3 briefly introduces the target and limit reference points laid out in Annex II of the FSA. Hence, the CCRF does not itself present any new provisions or operationalization of the precautionary approach, as compared with the FSA.

The FAO has followed up the CCRF with a host of supporting documents, including plans of action and technical guidelines. The FAO Technical Guidelines for Responsible Fisheries No. 2 deals specifically with the precautionary approach.[31] There is a separate two-page chapter on the precautionary approach and the burden of proof (Chapter 1). It notes that discussions about the precautionary approach often revolve around the placement of the burden of proof (Para. 7) and that 'although the precautionary approach to fisheries may require cessation of fishing activities that have potentially serious adverse impacts, it does not imply that no fishing can take place until all potential impacts have been assessed and found to be negligible' (Para. 7(b)). Rather, the precautionary approach entails that fisheries are subject to prior authorization and that a management plan is in place that specifies objectives and how impacts of the fisheries are to be assessed (Para. 7(c)); and that decisions about the authorization of fishing activities shall also take into account the expected benefits of the activities (Para. 7(c)): thus, a clear dismissal of at least the stricter versions of the precautionary approach.

[29] Agreement to Promote Compliance with International Conservation and Management Measures by Fishing Vessels on the High Seas, Rome, FAO, 24 November 1993, U.N.T.S. Vol, 2221, p. 91.

[30] Other areas covered are, inter alia, aquaculture, fisheries research and coastal area development.

[31] Precautionary Approach to Capture Fisheries and Species Introductions, FAO Technical Guidelines for Responsible Fisheries, No. 2, FAO, Rome, 1996.

Then follows Chapter 3 on what the precautionary approach to fisheries management is meant to entail[32]: management according to the precautionary approach is defined as exercising 'prudent foresight to avoid unacceptable or undesirable situations' (Para. 19), establishing 'legal or social management frameworks for all fisheries [including] rules controlling access to fisheries (e.g. all boats must be licensed), data reporting requirements, and processes for planning and implementing more comprehensive fishery management' (Para. 20), and taking 'explicit consideration of undesirable and potentially unacceptable outcomes and [providing] contingency and other plans to avoid or mitigate such outcomes' (Para. 22). There are paragraphs on management planning (Para. 3.2), implementation, monitoring and enforcement (Para. 3.3), re-evaluation of management systems (Para. 3.4) and implementation guidelines (Para. 3.5). Notably, the guidelines repeat the FSA's and the CCRF's requirement that stock-specific reference points shall be produced, but there is no attempt to operationalize them further. In sum, the precautionary approach is reduced to fairly traditional advice on how to design a system for fisheries management.[33]

This move away from the stringent precautionary approach (or principle) of international environmental law is evident also in the academic literature on international fisheries management. Hewison holds that a precautionary approach to fisheries management should involve, inter alia, a minimum estimate of stock biomass, a scientifically based and tested management system, a decision-making procedure characterized by transparency and stakeholder participation, and mechanisms to ensure compliance among fishers.[34] Garcia identifies the following steps: improving participation of 'non-fishery users', improving decision-making procedures, and strengthening monitoring, control and surveillance as well as raising penalties to deterrent levels.[35] Hence, while the precautionary *principle* is associated with the 'hard line' regimen proposed for the management of highly polluting activities, the precautionary *approach*

[32] The remaining chapters concern precautionary approaches to fisheries research, fisheries technology and introduced species.

[33] Tellingly, the guidelines have been criticized for not containing any mechanism for risk assessment. See, e.g., Fenichel et al. (2008, p. 122).

[34] Hewison (1996, pp. 320–329).

[35] Garcia (1994, p. 120).

arguably refers to practical ways and sets of measures which are precautionary in nature but may lead to a more 'realistic' application in fisheries.[36]

There is also a more 'critical' literature on the topic. Charles opines that there has been too much focus on the quantitative aspects of the precautionary approach, such as calculating 'correct' catch limits and reference points, and too little on the qualitative, structural uncertainties of fisheries management.[37] Among the latter, he includes model uncertainty (on the empirical basis for constructing models for stock assessment), implementation uncertainty (on the production of and responses to regulations, including fisher compliance) and institutional uncertainty (on the more overarching adaptation of actors to management systems).[38] Gerrodette et al. argue that a general change in the burden of proof is necessary, but not enough to achieve sustainable fisheries management[39]: '[A] truly precautionary approach requires a broader philosophical outlook than seeing the oceans as simply providing exploitable resources.'[40] These two contributions are fairly typical of the non-natural science literature on the precautionary approach to fisheries management,[41] focusing on the shortcomings of a narrow reference point approach to management. Again, the precautionary approach becomes a new device incorporating rather common-sensical perspectives on responsible fisheries management.[42]

[36] Ibid., p. 104.

[37] Charles (2002, p. 683).

[38] Ibid., pp. 683–684.

[39] Gerrodette et al. (2002, p. 657).

[40] Ibid., p. 657.

[41] This is based on a search in the ISI Web of Science (www.webofknowledge.com) (accessed 21 May 2020).

[42] As to whether the precautionary approach has achieved customary status in the fisheries domain, there is little guidance available in case law or the literature. As one example of the latter, Schatz et al. (2019, p. 218) argue that since CCRF paragraphs 6.5 and 7.5.1 have been widely applied in Regional Fisheries Management Organizations (RFMOs), 'it is submitted that the precautionary approach has become established as a general principle of fish stocks management'.

Implementation, Target Behaviour and State Practice

International instruments are normally implemented at the national level through incorporation or transformation, and any vague or generic provisions must be operationalized. For the instruments to have the intended effects, mechanisms must be in place—at the international or national level, or both—to induce behavioural change in the targets of the policies in question, in fisheries management: the fishers or fishing companies. In Chapter 1, I noted Andresen et al.'s typology *output*, *outcome* and *impact*. *Output* here refers to the adaptation of legislation and other forms of state practice, *outcome* to influence on fisher behaviour, and *impact* to the effects on the fish stocks and ecosystem.[43] *Output* and *outcome* are related to various stages in the implementation process; *impact* is the effect of implementation. State practice is the behaviour or conduct of states. As noted by Wood and Sender, for an act to qualify as state practice, it must be 'of the state', that is: attributable to it.[44] It comprises the actions of all branches of government and is not limited to conduct vis-à-vis other states or subjects of international law. Conduct within the state is also state practice if it relates to matters of international law. This includes administrative regulations and how they are interpreted and applied in practice.[45]

In the literature on the implementation of international environmental agreements, the concept of implementation is applied differently to capture distinct stages in this process. Much of the classic literature here is written at the intersection between international law and international relations (political science), often published as edited volumes with contributions from lawyers as well as political scientists.[46] Here two main approaches are discernible, one 'narrow' and one 'broad'. Jacobson and Weiss take implementation to refer only to national legislative activities, whereas subsequent activities to effectuate the provisions of the international agreement and/or national legislation are understood in terms

[43] Andresen et al. (2012, pp. 7–8).

[44] Wood and Sender (2017, p. 1).

[45] Ibid., pp. 2–3.

[46] As mentioned in Chapter 1, two classics are Weiss and Jacobson (1998) and Victor et al. (1998a).

of compliance or non-compliance with the treaty.[47] Hence, implementation is seen as covering only the first step in the two-step process of ensuring compatibility between (i) international commitments and national legislation (*output* in Andresen et al.'s typology); and (ii) international commitments/national legislation and the behaviour of target groups (*outcome*). This leaves implementation as a rather stylized and technical phenomenon to study: the question is simply whether the provisions of the international agreement have been incorporated in national legislation or not. By contrast, List and Rittberger differentiate between two levels of implementation activities: the transformation of international agreements into national law takes place at the national *normative* level, whereas de facto implementation occurs through other forms of state practice and target-group behaviour, and takes place at the national *factual* level.[48] The two approaches also differ as to who is seen as a legitimate actor in implementation activities. In the narrow understanding, only states may engage in implementation, as the concept is limited to the transformation or incorporation of international commitments into national legislation. In the broader version, also non-state actors can contribute to implementation, as the term is understood as all activities deliberately and explicitly aimed at bringing the behaviour of target groups into line with the requirements of international agreements (and national legislation insofar as it reflects international commitments).[49] Typical non-state actors would be NGOs and international organizations, with their implementation actions including the provision of financial assistance and training of national authorities or target groups in how

[47] Jacobson and Weiss (1998, p. 4). It is unclear how exactly these authors see the second step assessed in terms of compliance and non-compliance. These concepts are normally used in studies of either individuals' compliance with national law (as studied in criminology) or states' compliance with international agreements (as studied in international law and international relations). Presumably, the authors see implementation as a matter for international law and international relations, leaving compliance to the criminologists.

[48] List and Rittberger (1998, pp. 72–76). See also Hønneland and Jørgensen (2003, pp. 29–31). The distinction between implementation at the normative and factual levels does not overlap completely with the *output/outcome* divide. The transformation and incorporation of international obligations in national law constitute implementation *at the normative level, and 'output'*. Influence on target behaviour is implementation *at the factual level, and 'outcome'*. Other forms of state practice that reflect international obligations are implementation *at the factual level, and 'output'*.

[49] Victor et al. (1998b, pp. 3–5).

to implement the agreement. Such implementation activities by non-state actors may be conducted at the request of the authorities, or merely with their consent, and sometimes even without their consent (or knowledge).[50]

Concepts of implementation are closely linked to those of effectiveness and compliance.[51] If international commitments are not followed through at the national level, the agreement in question will have limited effect, as international instruments contain rights and obligations for states and have no direct legal effect for non-state actors—that can be achieved only through national legislation. The effectiveness of an international regime is often connected to either the achievements of the stated objectives of the regime or the solution of the problems that led to the establishment of the regime.[52] Whereas effectiveness is sometimes seen as primarily related to compliance, Victor et al. suggest that the degree of implementation may be a more reliable measure of effectiveness than the degree of compliance.[53] Where commitments are less ambitious, states may achieve full compliance with the formal provisions of a given agreement with very little behavioural adaptation.[54] Compliance may also be accidental, whereas implementation is by definition instrumental.[55] Hence, while compliance measures the degree of compatibility between a legal norm and the behaviour of target groups, implementation—in the broad understanding of the term—measures changes in target-group behaviour as the result of the introduction of the legal norm.

Notably, it is the broad version of implementation that is applied in this book. Both state practice and target behaviour are included here: the extent to which states have modified their practice (including legislation)

[50] For example, in Russia during the 1990s and early 2000s international NGOs such as Greenpeace and WWF were pivotal in implementing the state's international commitments under various biodiversity and nature-protection treaties, as the national bureaucracy lacked finances, capacity and competence. This was done by formal or informal agreement or by explicit or implicit consent (Jørgensen and Hønneland 2006, pp. 16–18).

[51] This paragraph builds on Hønneland and Jørgensen (2003, p. 30).

[52] Jacobson and Weiss (1998, p. 4).

[53] Victor et al. (1998b, p. 7). See discussion of Jacobson and Weiss' (1998) conception of the relationship between implementation and compliance above.

[54] Victor et al. (1998b, p. 7).

[55] Ibid., p. 7.

and fishers their behaviour as the result of MSC certification. By implication, it is acknowledged that implementation activities may be performed by non-state actors. Here it can be hypothesized that states adapt their fisheries regulations in order to ensure that their fishers become and remain MSC-certified, but also that fishers take direct action to achieve the same goal, by responding directly to MSC requirements without waiting for those requirements to be incorporated in national law. Both types of activities are covered in the broad understanding of implementation as long as the private certification standard itself contains provisions that, if complied with, would lead to the intended outcomes of the treaty in question. As an example of the former, there would normally have to be biological reference points in place for a fishery to pass MSC assessment, and reference points must be established (or endorsed) by states. As an example of the latter, strategies to avoid damage to the wider ecosystem can be developed by fishers themselves, as through codes of conduct at company level. To the extent that the MSC Certification Scheme incorporates the provisions of international agreements, such as the FSA, both these types of activities—like establishing binding (public) reference points and voluntary (private) codes of conduct—are deemed 'implementation activities'. I sporadically use the term 'implementation mechanism' of the MSC as a private certification scheme. By 'mechanism' is meant the instrument through which implementation activities are initiated and carried out. The MSC Fisheries Standard as such is just a set of rules according to which fisheries are assessed. But if the procedural requirements (the FCP) are included, and the activities carried out in accordance with them (see Chapter 3), we may speak of a mechanism, not just a set of rules.

While primarily considered a predecessor of legally binding international rules, soft law (non-legally binding international instruments) is also conceived as a mechanism to assist in the implementation of legally binding treaties and decisions of intergovernmental organizations, such as regional fisheries management organizations, RFMOs.[56] The CCRF is typically seen as a soft law instrument that might further the implementation of an international treaty like the FSA (although, as we saw above, the CCRF does not actually contribute much in terms of operationalizing the

[56] Friedrich (2008, pp. 1540–1541, 1551–1554).

precautionary approach).[57] Friedrich notes that the CCRF has 'important integrative functions',[58] and that the FAO 'resorts to numerous subtle ways through which states are drawn into flexible and discursive learning processes that often trigger important paradigm shifts of domestic law and policies towards more sustainable practices'.[59] He further observes that there is a division of labour between treaty law on fisheries and the CCRF on the one hand, and supportive mechanisms such as FAO's work on the ground and input from governments and civil society, on the other—here he lists private certification among such mechanisms.[60] The result is a 'cascade of soft law norms' that support the implementation of binding norms and further responsible fisheries management more widely.[61]

Private certification schemes do not produce soft law—that is the remit of states, at the intergovernmental level. But such schemes may have some of the same effects as soft law, as described by Friedrich. This book seeks to pinpoint the role that private certification can have in the 'cascade of instruments' geared towards implementing the precautionary approach to fisheries management.

References

Andresen, Steinar, Elin Lerum Boasson and Geir Hønneland, 'An International Environmental Policy Takes Shape', in Steinar Andresen, Elin Lerum Boasson and Geir Hønneland (eds), *International Environmental Agreements: An Introduction*, Abingdon: Routledge, 2012.

Bodansky, Dan, 'Deconstructing the Precautionary Principle', in David D. Caron and Harry N. Scheiber (eds), *Bringing New Law to Ocean Waters*, Leiden: Martinus Nijhoff, 2004.

Charles, Anthony T., 'The Precautionary Approach and "Burden of Proof" Challenges in Fisheries Management', 70 *Bulletin of Marine Sciences* 2002, pp. 683–694.

Ebben, Thomas, 'The Implementation of the Precautionary Approach in International Fishery Law: A Move towards "Green" Fishery', 15 *New Zealand Journal of Environmental Law* 2011, pp. 113–146.

[57] Ibid., pp. 1540–1543, 1546–1549.

[58] Ibid., p. 1541.

[59] Ibid., p. 1540.

[60] Ibid., p. 1559.

[61] Ibid., p. 1550.

Fenichel, Eli P., Jean Tsao, Michael L. Jones and Graham J. Hickling, 'Real Options for Precautionary Fisheries Management', 9 *Fish and Fisheries* 2008, pp. 121–137.

Freestone, David and Ellen Hey, 'Origins and Development of the Precautionary Principle', in David Freestone and Ellen Hey (eds), *The Precautionary Principle and Environmental Law: The Challenge of Implementation*, The Hague: Kluwer Law International, 1996.

Friedrich, Jürgen, 'Legal Challenges of Nonbinding Instruments: The Case of the FAO Code of Conduct for Responsible Fisheries', 9 *German Law Journal* 2008, pp. 1539–1564.

Garcia, Serge Michel, 'The Precautionary Principle: Its Implications in Capture Fisheries Management', 22 *Ocean & Coastal Management* 1994, pp. 99–125.

Gerrodette, Tim, Paul K. Dayton, Seth Macinko and Michael J. Fogarty, 'Precautionary Management of Marine Fisheries: Moving Beyond Burden of Proof', 70 *Bulletin of Marine Sciences* 2002, pp. 657–668.

Gillespie, Alexander, 'The Precautionary Principle in the 21st Century: A Case Study of Noise Pollution', 22 *International Journal of Marine and Coastal Law* 2007, pp. 61–87.

Hewison, Grant J., 'The Precautionary Approach to Fisheries Management: An Environmental Perspective', 11 *International Journal of Marine & Coastal Law* 1996, pp. 301–332.

Hønneland, Geir, *Russian Fisheries Management: The Precautionary Approach in Theory and Practice*, Leiden: Martinus Nijhoff, 2004.

Hønneland, Geir and Anne-Kristin Jørgensen, *Implementing International Environmental Agreements in Russia*, Manchester: Manchester University Press, 2003.

Jacobson, Harold K. and Edith Brown Weiss, 'A Framework for Analysis', in Edith Brown Weiss and Harold K. Jacobson (eds), *Engaging Countries: Strengthening Compliance with International Environmental Accords*, Cambridge, MA: MIT Press, 1998.

Jørgensen, Jørgen Holten and Geir Hønneland, 'Implementing Global Nature Protection Agreements in Russia', 9 *Journal of International Wildlife Law & Policy* 2006, pp. 33–53.

Kaye, Stuart B., *International Fisheries Management*, The Hague: Kluwer Law International, 2000.

List, Martin and Volker Rittberger, 'The Role of Intergovernmental Organizations in the Formation and Evolution of International Environmental Regimes', in Arild Underdal (ed.), *The Politics of International Environmental Management*, Dordrecht: Kluwer Academic, 1998.

Sands, Philippe and Jaqueline Peel with Adriana Febra and Ruth MacKenzie, *Principles of International Environmental Law*, fourth edition, Cambridge: Cambridge University Press, 2018.

Schatz, Valentin J., Alexander Proelss and Nengye Liu, 'The 2018 Agreement to Prevent Unregulated High Seas Fisheries in the Central Arctic Ocean: A Critical Analysis', 34 *International Journal of Marine and Coastal Law* 2019, pp. 195–244.

Victor, David G., Kal Raustiala and Eugene B. Skolnikoff (eds), *The Implementation and Effectiveness of International Environmental Commitments: Theory and Practice*, Cambridge, MA: MIT Press, 1998a.

Victor, David G., Kal Raustiala and Eugene B. Skolnikoff, 'Introduction and Overview', in David G. Victor, Kal Raustiala and Eugene B. Skolnikoff (eds), *The Implementation and Effectiveness of International Environmental Commitments: Theory and Practice*, Cambridge, MA: MIT Press, 1998b.

Weiss, Edith Brown and Harold K. Jacobson (eds), *Engaging Countries. Strengthening Compliance with International Environmental Accords*, Cambridge, MA: MIT Press, 1998.

Wood, Sir Michael and Omri Sender, 'State Practice', *Max Planck Encyclopedias of International Law* (Module: Max Planck Encyclopedia of Public International Law), last updated January 2017.

CHAPTER 3

The MSC Certification Scheme and the Precautionary Approach

Abstract The MSC Fisheries Certification Scheme, i.e. the Fisheries Standard and the Fisheries Process Requirements, includes both substantive and procedural requirements to precaution. First, the application of the approach is evaluated as such for fisheries under assessment. Second, it serves as a general interpretation device in the application of the Standard. Third, the Standard is designed to assist in the operationalization and application of precautionary biological reference points. Fourth, certification is not only a question of the status of the (target) stocks. The scheme involves a thorough assessment of the fishery's impacts on the marine ecosystem, including bycatch species, endangered, threatened and protected species (ETP species) and habitats—as well as management components such as a well-functioning legislative base and institutional set-up, transparency and stakeholder participation, compliance and enforcement, and review of the management system.

Keywords Marine Stewardship Council · MSC Fisheries Standard · MSC Fisheries Process Requirements

The main actors in an MSC assessment are the Conformity Assessment Body (CAB), the MSC itself, the accreditation body Assurance Services

© The Author(s), under exclusive license to Springer Nature Switzerland AG 2021

G. Hønneland, *Blue Governance in the Arctic and Antarctic*, Palgrave Studies in Maritime Politics and Security,
https://doi.org/10.1007/978-3-030-72585-3_3

International (ASI) and the fishery client seeking certification. The MSC is the scheme owner—it produces the standards and issues certificates, but is not directly involved in the assessments except for providing technical reviews of assessment reports. The MSC is a non-governmental, non-profit organization headquartered in London, with regional offices in the USA and Australia and local representations in more than 20 countries. The assessments are performed by certification bodies, CABs, which compete for assignments among fishery clients on a commercial basis.[1] The MSC has appointed the ASI as its accreditation body. CABs must be accredited by the ASI in order to perform MSC assessments, and they are under constant ASI scrutiny through document review and physical inspection. The 'fishery client' is anyone applying for certification for one or more fishing vessels—it may be a company, a regional or national association, or a group of companies or associations from several countries.

The MSC has three main types of programme documents, here referred to collectively as the Certification Scheme: standards (containing substantive requirements for certification); process requirements (to assessments according to the standards); and guidance documents (on how the standards and process requirements are to be interpreted). There are standards pertaining to fisheries, traceability of fish products ('Chain of Custody'), and seaweeds (jointly with the Aquaculture Stewardship Council (ASC)), each with supporting process requirements and guidance documents. The documents are revised in five-year cycles.

In the following, the MSC assessment process is first presented, then the substantive requirements of the MSC Fisheries Standard, and finally the integration of the precautionary approach in the Standard.[2] As the Standard with Guidance comprises close to 300 pages and the FCP with Guidance nearly 200 pages, this presentation can only scrape the surface of the MSC processual and substantive requirements.

[1] Some of the largest CABs are global certification and classification societies such as Control Union, DNV GL and Lloyd's Register.

[2] The versions referred to are the MSC Fisheries Standard and Guidance v2.01 and the MSC Fisheries Certification Process and Guidance v2.1, both of 31 August 2018. All MSC programme documents are downloadable at https://www.msc.org/for-business/certification-bodies/fisheries-standard-program-documents (accessed 24 August 2020).

Procedural Requirements: The MSC Fisheries Certification Process

The first step in an assessment is to confirm that the fishery is within scope for MSC certification (FCP 7.4). A fishery is within scope if the target species are not amphibians, reptiles, birds or mammals (FCP 7.4.2.1) and poisons or explosives are not used (FCP 7.4.2.2). Further, the fishery must not be conducted under a controversial unilateral exemption to an international agreement (FCP 7.4.3), be embroiled in dispute, or lack a mechanism for solving disputes (FCP 7.4.5). Nor may it include entities that have been successfully prosecuted for forced or child labour (FCP 7.4.4). Enhanced fisheries, where the growth of the fish is assisted at some stage of its development, as well as introduced species, are within scope only under specific circumstances (FCP 7.4.6, 7.4.7).

The next step is to define the Unit of Assessment (UoA), which is to include the target stock(s); the fishing method or gear type(s), vessel type(s) and/or practices; and the group of vessels whose fishing operations are to be covered by the assessment (FCP 7.5.2). Later in the process, the client—the company or group of companies seeking certification—decide on the Unit of Certification (UoC) (FCP 7.5.3): whether the whole UoA or just a specific group of vessels shall be covered by the specific certificate. Other vessels in the UoA whose fishing activities have been covered by the assessment are termed 'other eligible fishers' and can join the certificate through a sharing agreement with the client (FCP 7.5.7). This is normally not done free of charge, as the certification process involves considerable costs for the client.

The public announcement of an assessment takes place with the posting on the MSC website of the Announcement Comment Draft Report (ACDR) (FCP 7.15.1). The ACDR is an almost full version of the assessment report, but with indicative scoring ranges rather than specific scores (see next section). The assessment team, three independent experts working on behalf of the CAB, draws on all available evidence, including scientific literature, to evaluate the fishery. The announcement of the fishery is followed by a 60-days' consultation period (30 days for re-assessments), during which stakeholders are encouraged to provide comments to the ACDR or other inputs relevant to the assessment. All written stakeholder submissions are published on the MSC website as well as in subsequent versions of the assessment report, along with the response of the assessment team.

The assessment process involves a site visit, where the assessment team interviews stakeholders in the fishery, like scientists, managers, representatives of enforcement bodies, industry groups and NGOs. Next, the team holds a scoring meeting, where the balance of the evidence collected until that point is weighed up (FCP 7.17.2). The fishery is scored according to a fine-meshed system of Scoring Issues (SIs)—'the Assessment Tree' (see next section)—attached to various Performance Indicators (PIs), within the three MSC Principles: Principle 1 (P1) on the status of the target stock(s), Principle 2 (P2) on the ecosystem impact of the fishery and Principle 3 (P3) on the effectiveness of the management system. Specific requirements are assigned to each SI: the Scoring Guideposts (SGs) for scores at 60, 80 and 100. For a fishery to pass the assessment, no SI may score less than 60, and the average weighted score of each of the three Principles must be at 80 or above. Most PIs consist of several SIs, and scores are given at increments of five points (FCP 7.17.5).

If a score between 60 and 80 is given for an individual SI, one or more 'auditable and verifiable conditions [for certification]' must be set by the assessment team (FCP 7.18.1). The team shall draft conditions that, when implemented, are to result in improved performance to at least the 80 level within the five-year period during which the certificate is valid (FPC 7.18.1.3). The conditions are to include milestones that the client must meet at each annual surveillance audit (see below) during the five-year certification period. The milestones shall, inter alia, identify 'measurable improvements and outcomes (using quantitative metrics) expected each year' (FCP 7.18.1.4.a). Once the CAB has determined the conditions and milestones to be attached to the fishery and has taken into account all available information as per the last day of the site visit, the assessment team completes its Client and Peer Review Draft Report (CPRDR) (FCP 7.19.1). The CPRDR is sent to the client for comments and production of a Client Action Plan (CAP), detailing how the client intends to work towards meeting the annual milestones in the conditions set by the assessment team in order to bring the score up to 80 by the end of the certification period. The CPRDR also goes to the MSC Peer Review College (FCP 7.19.1), where two suitable peer reviewers are drawn from a pool of qualified experts. These reviewers fill in a detailed sheet that includes comments on every PI in the Standard, which, along with the assessment team's response, is included in all subsequent versions of the assessment report.

When the assessment team has responded to the comments from the client and the peer reviewers, the Public Comment Draft Report (PCDR) is posted on the MSC website for public comments during a 30-day period (FCP 7.20). The PCDR also goes back to the peer reviewers for a second round of comments (FCP 7.20.9) and to the MSC for 'Technical Oversight' (FCP 7.20.10). The latter involves a 'legality check' in which the scoring of each SI is controlled, considering the wording of the guideposts, the relevant guidance and interpretations as well as the assessment team's justification and documentation of its scores. When the team has responded to the comments from stakeholders, peer reviewers (second round) and the MSC, the Final Draft Report is produced and published on the MSC website (FCP 7.22), followed by an objection period of 15 working days (FCP 7.23).

The objection procedure is outlined in a separate annex to the FCP (Annex PD). Importantly, the procedure is not intended to result in a new assessment of the fishery against the Standard, but to determine whether the assessment team made a procedural error (including using the available evidence incorrectly in determining the scores) that is 'material to the determination or the fairness of the assessment' (PD 2.1.1.1). The objection procedure is open only to parties who made written submissions earlier in the assessment process (PD 2.1.2) and is led by an Independent Adjudicator (a civilian judge, attorney or professor of law) appointed by the MSC for a three-year period (PD 2.2). The Independent Adjudicator must first determine whether the filed objection is in the required form and that it has reasonable prospects of success (PD 2.4.1); further, that it is not spurious or vexatious (PD 2.4.2a), in which case it is dismissed. Then follows a 15-day period for consultations between the objector and the CAB, organized by the Independent Adjudicator (PD 2.5). If agreement is not reached, 'written representation' commences, with defined procedures for correspondence over a 30-day period (PD 2.6). If an agreement is still not reached, the case goes to formal adjudication (PD 2.7), including an oral hearing. The decision of the Independent Adjudicator is final (PD 2.7–2.8).

The MSC certificate is valid for five years. The state of the fishery, as well as progress towards the set milestones for any conditions attached to the certificate, is monitored by the assessment team at annual surveillance audits (FCP 7.28). If a fishery client is behind target at a surveillance audit, remedial action is defined. If the fishery is not back on track for the next annual surveillance audit, it is suspended from MSC certification.

A corrective action plan must be produced within 90 days of suspension; if the terms of the plan are not complied with in the set timeframe, the certificate is withdrawn (FCP 7.28.16.2.b). If fishery clients wish to remain certified beyond the first five-year period, re-assessment must be commenced no later than 90 days after the fourth anniversary of the certificate (FCP 7.30).

Substantive Requirements: The MSC Fisheries Standard

The MSC Fisheries Standard is organized in the Assessment Tree, which spells out the specific requirements (guideposts) against which a fishery is assessed: 89 SIs spread over 28 PIs within the three MSC Principles. (There is one level between the Principle and the PIs, the Components.) The principles are thematically defined, but the PIs can also be grouped into outcome, management and information indicators. For outcome indicators, it is required that the fish stocks and other components of the ecosystem are at acceptable levels; for management indicators that adequate management measures are in place; and for information indicators that there exists sufficient information to enable appropriate management decisions.

Principle 1 (P1) is defined as follows:

> **Principle 1: Sustainable target fish stocks**
>
> A fishery must be conducted in a manner that does not lead to over-fishing or depletion of the exploited populations and, for those populations that are depleted, the fishery must be conducted in a manner that demonstrably leads to their recovery. (MSC Fisheries Standard, p. 5)

As shown in Fig. 3.1, P1 consists of one outcome and one management component; several PIs also include elements of information indicators. Component 1 has only two PIs: one on the status of the stock (PI 1.1.1) and one on stock rebuilding (PI 1.1.2). In order to pass PI 1.1.1 (i.e. to achieve a score of 60), it must be 'likely' (defined as at least 70% probability) that the target stock is above the point where recruitment would be impaired (PRI). To pass without conditions (i.e. to score 80 or above), it must be *highly* likely (80% probability) that the stock is above PRI, and the stock must be fluctuating around a level consistent with MSY. To achieve a score of 100, there must be a high degree of certainty

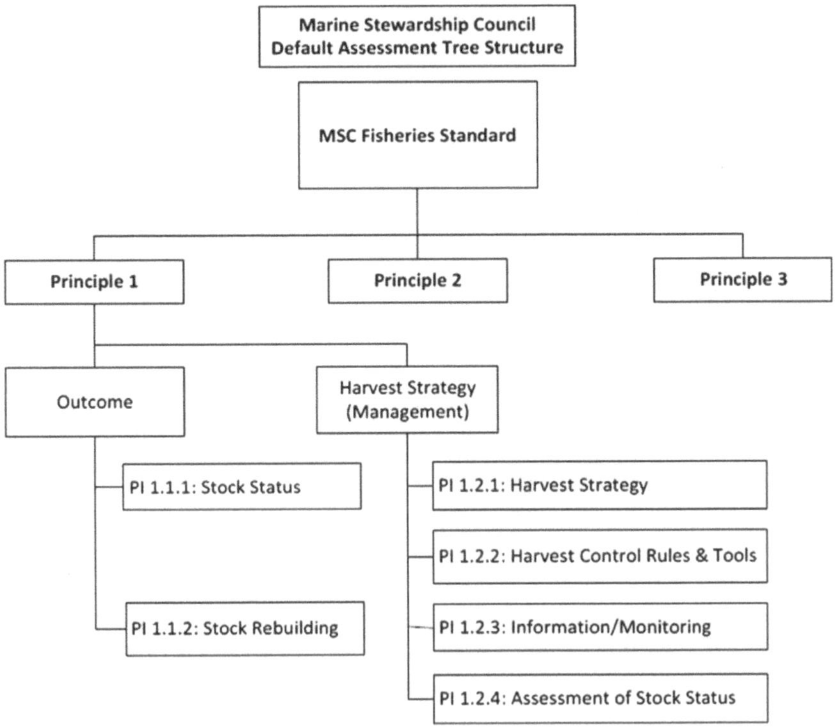

Fig. 3.1 Principle 1 Assessment Tree

(95% probability) that the stock is above PRI and that it has been fluctuating around or been above MSY over recent years. PI 1.1.2 on stock rebuilding is scored only if SG 80 is not met for PI 1.1.1. In order to pass PI 1.1.2, a rebuilding timeframe must have been specified for the stock that is 20 years or two times the stock's generation time (whichever of the two is shorter), and monitoring must be in place to determine whether the rebuilding strategy is effective. To pass unconditionally, there must be evidence that the strategy is working.

Component 2 of P1 comprises PIs related to the existence of a harvest strategy (PI 1.2.1), a harvest control rule (HCR) (PI 1.2.2), information to support the harvest strategy (PI 1.2.3) and scientific stock assessments (PI 1.2.4). An HCR is defined as 'a set of well-defined pre-agreed rules or actions used for determining a management action in response to changes in indicators of stock status with respect to reference points'.[3] For a fishery to pass PI 1.2.2, there must be a generally understood HCR expected to reduce the exploitation rate if PRI is approached, and there must be some evidence that it is appropriate for the fishery under assessment.[4] To pass unconditionally, a well-defined HCR must be in place capable of keeping the stock at MSY, supported by evidence. To achieve a 100 score, also the ecological role of the stock must be taken into account; the HCR must consider a wide range of uncertainties, and evidence must show *clearly* that the HCR is working.

Principle 2 (P2) is defined as follows:

> **Principle 2: Environmental impact of fishing**
> Fishing operations should allow for the maintenance of the structure, productivity, function and diversity of the ecosystem (including habitat and associated, dependent and ecologically related species) on which the fishery depends. (MSC Fisheries Standard, p. 5)

The P2 Assessment Tree (see Fig. 3.2) has five thematically defined Components: on primary species; secondary species; endangered, threatened and protected (ETP) species; habitats and the wider ecosystem. Primary and secondary species concern bycatch—the former of species that are managed by biological reference points, the latter of species that are not.[5] Each Component is split into PIs on outcome, management and information. P2 is clearly the most complex of the three MSC Fisheries Standard Principles. Not only does it contain a higher number of SIs than P1 and P3 taken together (49, as opposed to 21 and 19, respectively); it also has a more comprehensive set of guidance for interpretation

[3] MSC-MSCI Vocabulary, Version 1.2, 28 March 2019, available at https://www.msc.org/docs/default-source/default-document-library/for-business/program-documents/chain-of-custody-supporting-documents/msc-msci_vocabulary_v1-2.pdf?sfvrsn=cef284dd_14 (accessed 22 August 2020).

[4] The Guidance specifies that 'well-defined' means 'in some written form that has been agreed by the management agency, ideally with stakeholders', clearly stating 'what actions will be taken at what specific trigger reference point levels' (GSA2.5). 'Generally understood' implies that HCRs 'have been applied in some way in the past, but have not been explicitly defined or agreed' (ibid.).

[5] Hence, the absolute requirement to have reference points in place applies only to the target stock(s), assessed under P1.

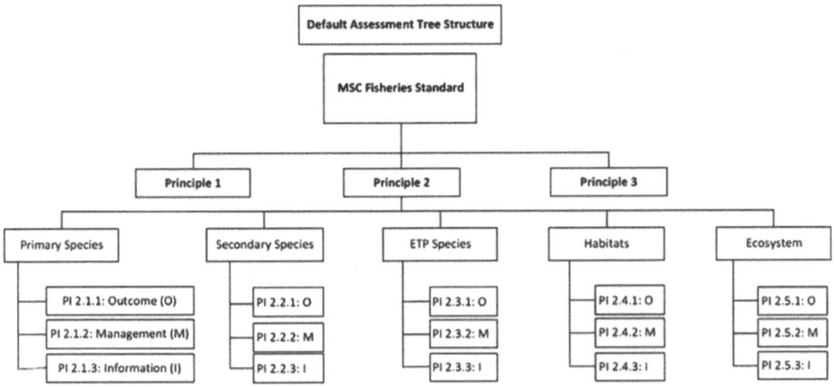

Fig. 3.2 Principle 2 Assessment Tree

and scoring. Further, assessment teams must assess not only the environmental impacts of the client fishery (the UoA), but also the accumulative impact of all MSC fisheries in the same region.

Due to the lack of space to present all the PIs under P2, let us take the outcome PI for ETP species as an example. In order to pass PI 2.3.1, the effects of the UoA fishery on ETP species must be known, and must be within the limits set by national legislation and international agreements, if such exist. Further, known direct effects of the fishery must be likely not to hinder recovery of the ETP species. For the fishery to pass without condition, the combined effects of all MSC-certified fisheries in the region must be known and *highly* likely to be within the limits set by national legislation and international agreements. Also, the direct effects of the fishery must be highly likely not to hinder the recovery of ETP species, and *indirect* effects of the fishery must be considered highly likely not to create unacceptable impacts. (The guidepost for a 100 score further strengthens the requirement to 'a high degree of certainty' and also expands the material requirements.) The outcome PIs for habitats (PI 2.4.1) and the wider ecosystem (PI 2.5.1) follow the same logic.

Principle 3 (P3) is defined as follows:

Principle 3: Effective management

The fishery is subject to an effective management system that respects local, national and international laws and standards and incorporates institutional and operational frameworks that require use of the resource to be responsible and sustainable. (MSC Fisheries Standard, p. 5)

P3 is split into two Components (see Fig. 3.3): one on the wider management framework of the fishery (at national and international levels) and one on the fishery-specific management system, i.e. the system directly responsible for the management of the UoA fishery. Whereas P1 and P2 focus mainly on the status of ecosystem components and the appropriateness of specific management measures (whether they have, or are likely to have, the intended effects), P3 concerns *structure* and *process*, e.g. whether appropriate legislation (PI 3.1.1), dispute-resolution mechanisms (PI 3.1.1), opportunities for industry and other stakeholders to get involved in the management process (PI 3.1.2), enforcement mechanisms (PI 3.2.3) and appropriate objectives for the fishery (PIs 3.1.3 and 3.2.1) are in place. For example, for PI 3.2.4 on evaluation of the management system, there must at least be occasional internal reviews of the system for the fishery to pass, regular internal and occasional external reviews for the fishery to pass without condition, and regular internal and external reviews for the score to reach 100.

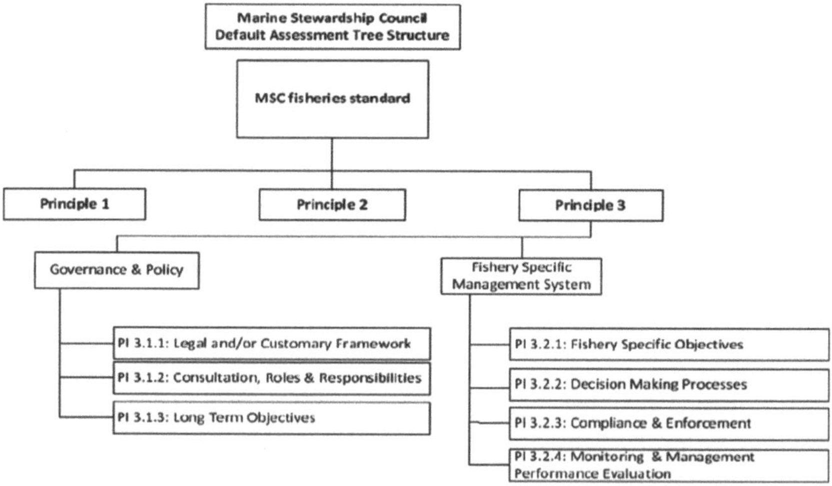

Fig. 3.3 Principle 3 Assessment Tree

The Precautionary Approach in the MSC Certification Scheme

The precautionary approach to fisheries management is incorporated at various levels in the MSC Certification Scheme. First, the application of the precautionary approach in the fishery under assessment is scored as such twice under P3, namely under PI 3.1.3 on the overarching management structure and under PI 3.2.2 on the fishery-specific management system. Objectives consistent with the precautionary approach must under PI 3.1.3 be either implicit or explicit in the management regime for the fishery to pass/pass without a condition, respectively. To score 100, such objectives must be required by management policy. The Standard refers to the definition of the precautionary approach in the FSA, Article 6 (SA4.5), as to its interpretation. The Guidance, however, notes that it is not the intention 'to point [assessment] teams towards Article 6, Annex II of the Fish Stocks Agreement for a prescriptive list of what must appear in management policy in relation to the precautionary approach' (GSA4.5). Nor is it meant as 'a second opportunity to score UoAs on the use or otherwise of target and limit reference points which are scored under PI of the [assessment] tree' (ibid.). Rather, '[t]he emphasis of this PI is on the presence or absence of long term objectives of the management agency for all UoAs under its control' (ibid.). Further, SI 3.2.2(a) requires that 'decision-making processes use the precautionary approach and are based on best available information' (SA4.8).[6]

Second, the precautionary approach is a general interpretation device in the application of the Standard as such. Under 'General Requirements' in the Guidance to the Standard (GSA1.1), the first of two sections concerns the precautionary approach[7]: 'In the MSC standard the application of the precautionary approach in fisheries management systems is explicitly scored in PIs 3.1.3 and 3.2.2. However, the MSC also intends the precautionary approach to be applied implicitly throughout the Certification Requirements' (GSA1.1). This procedural requirement was confirmed by

[6]This SI, like a few other SIs in the Assessment Tree, exists only at SG 80, which implies that the fishery cannot fail on this SI (as there is no SG 60); nor is there the possibility for it to achieve a 100 score that can balance out the average score at 80 if another SI has achieved only a 60 score.

[7]The other one concerns the MSC's intent and understanding of the standard in relation to IUU fishing.

Independent Adjudicator John McKendrick QC in the Murmanseld 2 objection case (see Chapter 4): 'When interpreting the [MSC Standard] I should assume an interpretation that implicitly gives effect to the precautionary approach as opposed to not doing so is more likely to be correct.'[8] Here we see the precautionary approach applied at different levels. First, the application of the approach in the fisheries under assessment is evaluated. Second, precaution is introduced as a device for the MSC itself in designing the Standard, and arguably also for assessment teams when applying it.

Third, and most importantly, the MSC Fisheries Standard is designed to assist in the operationalization and application of the precautionary biological reference points envisaged by Article 6 and Annex II of the FSA. Previous versions of the Standard had a separate PI under P1 on the existence of reference points. This has now been removed, and requirements to apply reference points are found primarily in the management Component of P1, in particular under PI 1.2.2 on HCRs. As noted in Chapter 2, reference points are normally set for the size of the spawning stock and for fishing mortality, with one target reference point (to aim for) and one limit reference point (to trigger new management action). The key purpose of an HCR is to trigger (precautionary) action if the spawning stock falls below a certain level, or fishing mortality rises above a certain level.

Finally, fisheries sustainability certification is not only about the status of the (target) fish stocks. Reflecting developments in the law of the sea and international ocean politics more widely, the ecosystem effects of fisheries have arguably become the main topic in MSC assessments, with new versions of the Standard introducing increasingly stricter requirements. In effect, the entire P2 reflects an ecosystem approach to fisheries management, with its focus on the target fishery's impacts on bycatch, ETP species, habitats and the wider ecosystem. Especially with the obligation to assess the cumulative ecosystem effects of all MSC-certified fisheries in a region, P2 assessment is a far more comprehensive exercise than assessing P1 or P3. Hence, there are both substantive and procedural requirements to precaution in the MSC Certification Scheme, and a link between the two: the more precautionary the Standard and the

[8] Murmanseld 2 1-FA, p. 298 (see reference lists for Chapters 4 and 5 for reference technique for MSC assessment reports).

scoring practices, the less danger that lack of precaution in the fisheries management system will be overlooked.

References

MSC Fisheries Certification Process and Guidance v2.1, 31 August 2018.
MSC Fisheries Standard and Guidance v2.01, 31 August 2018.

CHAPTER 4

MSC Certification of Antarctic and Arctic Fisheries

Abstract This chapter examines the results of MSC certification in four clusters of fisheries in polar waters: one in the Antarctic and three in the (sub-)Arctic. MSC certification has had several precautionary effects in these fisheries. It has (i) been an attestation to the appropriateness and soundness of the established reference points and harvest control rules (HCR) in the Barents Sea cod and haddock fisheries; (ii) helped to ascertain the appropriateness and soundness of HCRs in the Southern Ocean toothfish and krill fisheries and (iii) been the direct cause of the development and establishment of reference points and HCRs in the Arctic lumpfish fisheries. Improvements have been made to management systems beyond the introduction of reference points and HCRs: new research and information gathering has been conducted in the Southern Ocean, and efforts have been made to reduce impacts on bottom habitats in the Barents Sea and on seabirds and marine mammals in the Arctic lumpfish fisheries. Effects are more modest in the Norwegian Sea mackerel fisheries, where the coastal states for several years have not been able to agree on sustainable fishing quotas. MSC certification has not alleviated the situation, and in 2020 the mackerel certificate was suspended.

Keywords Southern Ocean · Barents Sea · Northeast Atlantic · Ecolabelling · Marine Stewardship Council

© The Author(s), under exclusive license to Springer Nature Switzerland AG 2021

G. Hønneland, *Blue Governance in the Arctic and Antarctic*, Palgrave Studies in Maritime Politics and Security, https://doi.org/10.1007/978-3-030-72585-3_4

We now move from analysis of the MSC Standard itself to study how it has been applied in the Southern Ocean, the Barents Sea, the Norwegian Sea (internationally managed fisheries), and in waters under national jurisdiction in Greenland, Iceland and Norway. Whereas the next chapter presents the effects of certification on fisher and management behaviour, this chapter details the processes and assessment results of MSC assessments. What potentially dictates behavioural adaptation are the improvement conditions attached to MSC certificates: the specific requirements for improvements that must be achieved during the five years the certificates are valid. Conditions follow from the expert opinion of the assessment teams appointed by the CABs (one expert per MSC Principle), but are influenced by input from peer reviewers, stakeholders and, where relevant, objectors to certification and, in the final instance, the MSC's Independent Adjudicator.

All information in the running text below refers to Tables 4.1, 4.2, and 4.3, one for each cluster of fisheries. That information is taken from the assessment reports for each fishery named in the reference list. It follows from the context which report is the source for each piece of information[1]; page reference to the primary source is provided only for direct citations,[2] using the coding provided in the footnote to the reference list (e.g., 1-FA = 1st full assessment report; 2-SA = 2nd annual surveillance audit report; and 3-RA = 3rd re-assessment report).

Krill and Toothfish in the Southern Ocean

The South Georgia Patagonian toothfish fishery (with the Government of South Georgia and the South Sandwich Islands, a British Overseas Territory, as client) applied for MSC certification in 2002 (see Table 4.1). The New Zealand Ross Sea toothfish fishery and the Norwegian Aker Biomarine krill fishery (both with individual companies as clients) applied in 2007 and 2008, respectively. All three assessments took a long time to

[1] For example, all information about the Aker Biomarine krill assessment is taken from the (re-)assessment reports for that fishery listed in the reference list. Information on fulfilment of previous conditions is taken from the surveillance audit report of the preceding certification cycle.

[2] It is impossible within the limited space available for this book to provide detailed page reference for every piece of information compiled in the tables. Page references related to conditions are provided in Chapter 5.

Table 4.1 Assessment and re-assessment results of krill and toothfish fisheries in the Southern Ocean re-certified as per 2020, compiled by the author based on assessment reports available on the MSC website (see reference list)

Fishery Initial assessment*	*Announced/certified*	*Conditions*	*Stakeholder submissions***	*Objections*
South Georgia Patagonian toothfish (South Georgia and the South Sandwich Islands/UK)	February 2001/1 March 2004***	– Nine conditions (different numbering than today), main focus on confirmation of stock identity; impacts on dependent species; lacking research on ecosystem relations of toothfish, incl. with benthic habitats; IUU	– Birdlife International, Scientific Committee on Oceanic Research (SCOR), The Antarctica Project, FAO, WWF, Falkland Conservation, Greenpeace International, National Audubon Society (issues of concern largely scientific uncertainties regarding the stock and ecosystem effects)	– National Environmental Trust and The Antarctica Project (rationales (selection): species identification; stock modelling; impacts on dependent/associated species; compliance with UNCLOS and application of MSY; certification in context of widespread IUU fishing) (certification determination remanded to the CAB for re-scoring of some PIs)

(continued)

Table 4.1 (continued)

*Fishery** *Initial assessment*	*Announced/certified*	*Conditions*	*Stakeholder submissions***	*Objections*
Ross Sea toothfish (New Zealand)	21 November 2007/16 November 2010	– Eight conditions (different numbering than today), main focus on lack of research to locate areas of complex benthic habitats; lack of research on major predators and prey of toothfish; lack of assessment of the risk posed by the fishery to populations of significant bycatch	– Centre for Biological Diversity (CBD), ASOC, WWF, Argos Georgia Ltd. (the fishery client), Friends of the Ross Sea Ecosystem, Mr Joseph Eastman, Environment and Conservation Organisations of NZ, Greenpeace Int'l, Mr Donald Siniff (general lack of information)	– ASOC (except WWF) (rationales: a wide range of procedural and substantive issues, e.g. lack of fundamental data about the fishery; uncertainty regarding the distinction of the stock; the toothfish's vulnerability to depletion; and IUU fishing) (objection to the scores against 36 PIs, four of which were upheld by the Independent Adjudicator)
Aker Biomarine krill (Norway)	18 September 2008/15 June 2010	– PIs 1.1.2, 2.1.1, 2.1.2, 2.5.3	– Pew coordinated science response'****; ASOC, Pew's Antarctic Krill Conservation Project (lack of information on krill population dynamics and ecosystem effects)	– ASOC (except WWF) (main rationale: information is insufficient to ensure that neither krill populations nor predators are harmed by fishing pressure) (certification determination remanded to the CAB for re-scoring of some PIs)
SARPC toothfish (Réunion)	1 October 2009/23 August 2013	– PIs 1.2.4, 2.1.2, 2.1.3, 2.3.1, 3.2.1, 3.2.2	– WWF-Australia, ASOC, Ligue pour la Protection des Oiseaux (P1 and P2 in general)	None

*Fishery**	*Announced/certified*	*Conditions*	*Stakeholder submissions***	*Objections*
Initial assessment				
Heard Island and McDonald Islands toothfish (Australia)	23 September 2010/16 March 2012	– PIs 1.2.1, 1.2.2, 1.2.2 (HCR), 1.2.4, 2.3.4, 3.1.2	– WWF-Australia, Nigel Brothers Marine Ecology & Technology (stock assessment, habitats, IUU)	None
Re-assessment				
South Georgia Patagonian toothfish	23 July 2013/15 September 2014	None	– Argentinian Embassy London (political grounds)	None
Aker Biomarine krill	19 November 2013/6 November 2014	None	– WWF-Norway (outdated stock survey, impact on predators, anticipated expansion of fishery)	None
Ross Sea toothfish	19 August 2014/12 January 2016	– PI 1.2.4	– ASOC (inappropriate reference points, negative ecosystem impacts)	None
Heard Island and McDonald Islands toothfish	7 July 2016/11 July 2017	None	None	None
SARPC toothfish	11 January 2018/20 December 2018	– PI 2.2.2	– Mr Thibaut Thellier, Chargé des milieux marins des îles australes (technical clarifications)	None

*The official names of the fisheries in the MSC system have been shortened here to save space. Full names are given in the reference list

**Only submissions following the publication of the PCDR are included in the Table; many fisheries also received stakeholder submissions at earlier stages of the assessment. Technical oversight comments from MSC are not included, although formally they fall under stakeholder comments

***According to the MSC website, the fishery was announced 12 March 2002, but it follows from the report that work commenced in February 2001

****Norwegian College of Fishery Science, Monterey Bay Aquarium and New England Aquarium, group of scientists from Argentina, Canada, Mexico, Peru, Spain and the USA, referred to by the assessment team as 'Pew coordinated response'

Table 4.2 Assessment and re-assessment results of cod and haddock fisheries in the Barents Sea certified as per 2020, compiled by the author based on assessment reports available on the MSC website (see reference list)

*Fishery/client** *Initial assessment*	*Announced/certified*	*Conditions*	*Stakeholder submissions***	*Objections*
Norway (Norw. Seafood Council)	3 September 2008/26 April 2010	– PIs 2.1.1, 2.3.3, 2.4.1	– WWF-Norway (several comments across all three principles)*** – Directorate of Fisheries (non-material)	None
Ocean Trawlers (Russia)	24 December 2008/24 October 2010	– PIs 1.2.1, 1.2.2, 1.2.3, 2.1.1, 2.1.2, 2.4.1, 2.4.2, 3.1.2, 3.1.3	– WWF-Russia (several comments across all three principles)***	None
UK Fisheries/DFFU/Doggerbank	13 January 2011/3 May 2012	– PIs 2.4.1, 3.1.1, 3.1.2, 3.2.2, 3.2.5	None	None
Saint Malo/Euronor (France)	2 September 2010/19 April 2012	– PIs 2.4.1, 2.4.1	None	None
Faroe Islands/Iceland	28 June 2011/16 August 2012	None	None	None
FIUN (Russia)	22 March 2012/25 June 2013	– PIs 2.1.1, 2.1.2, 2.4.1, 2.4.2, 3.1.2	– American Bird Conservancy (impact on seabirds (only for longline))	None
AGARBA (Spain)	7 June 2012/28 November 2013	– PIs 2.1.3, 2.3.2–2.4.3	None	None

*Fishery/client**	*Announced/certified*	*Conditions*	*Stakeholder submissions***	*Objections*
Initial assessment				
Strelets/Eridan (Russia)	21 March 2013/6 May 2014	None	– WWF-Russia (PI 2.4.2)	None
Greenland	5 August 2013/5 May 2015	– PI 2.4.3	– WWF-Germany (PIs 2.4.1–2.4.3)	WWF-Germany (PIs 2.4.1–2.4.3)
Arkhangelsk Trawl Fleet (Russia)	24 October 2014/26 January 2016	– PIs 2.3.1, 2.3.2, 2.3.4, 2.4.2, 2.4.3	– WWF-Russia (PI 2.4.2)	None
Oceanprom (Russia)	7 December 2017/11 June 2019	– PIs 2.3.2, 2.3.3	None	None
Murmanseld 2 (Russia)	22 October 2018/5 March 2020	– PIs 2.3.2, 2.4.1, 2.4.2	– WWF-Germany and WWF-Russia (PIs 2.4.1, 2.4.2)	– WWF-Germany and WWF-Russia (PIs 2.4.1, 2.4.2) (partial acceptance)
Re-assessment				
Norway (Norw. Fishermen's Association)	16 December 2014/6 October 2015	– PIs 2.3.1, 2.4.1	– WWF-Norway (PIs 2.4.1–2.4.3)	None
Ocean Trawlers (Russia)	14 October 2014/20 September 2016	None	– WWF-Germany (PIs 2.4.1–2.4.3)	– WWF-Germany (PIs 2.4.1–2.4.3) (compromise achieved before oral hearing)
UK Fisheries/DFFU/Doggerbank	12 August 2016/14 November 2017	– PIs 2.3.2–2.4.3	– WWF-Germany (PI 2.4.1)	None
Faroe Islands/Iceland	8 September 2016/21 August 2017	None	None	None
Saint Malo/Euronor (France)	22 September 2016/13 October 2017	– PIs 2.4.1–2.4.3	None	None

(continued)

Table 4.2 (continued)

Re-assessment				
FIUN cod and haddock	20 July 2017/28 August 2018	– PI 2.4.2	None	None
Strelets/Eridan (Russia)	26 September 2017/2 August 2019	None	None	None
Greenland	29 September 2017/30 April 2019	None	None	None
AGARBA (Spain)	24 May 2018/26 April 2019	None	– WWF-Germany (PIs 2.4.1, 2.4.2)	None

*The official names of the fisheries in the MSC system and the clients have been shortened here to save space. Full names are given in the reference list

**Only submissions following the publication of the PCDR are included in the Table; many fisheries also received stakeholder submissions at earlier stages of the assessment. Technical oversight comments from MSC are not included, although formally they fall under stakeholder comments

***These submissions were not specific as to which PIs they were meant to address

Table 4.3 Assessment and re-assessment results of mackerel fisheries in the Northeast Atlantic certified as per 2020, compiled by the author based on assessment reports available on the MSC website (see reference list)

*Fishery/client** *Initial assessment*	*Announced/certified*	*Conditions*	*Stakeholder submissions***	*Objections*
Scottish Pelagic Sustainability Group	31 May 2007/21 January 2009	– Four conditions (different numbering than today), main focus on strategies to reduce slippage mortality and overfishing, and implement new harvest control rule	None	None
Norwegian Fishermen's Sales Organization for Pelagic Fish	31 January 2008/30 April 2009	– Four conditions (different numbering than today), main focus on recording of slippage and bycatches, and implementation of new harvest control rule	None	None
Pelagic Freezer Trawler Association (Netherlands)	7 February 2008/8 July 2009	– PIs 1.2.2, 2.3.2, 2.3.3	None	None
Irish Pelagic Sustainability Group	6 August 2008/24 August 2009	– PIs 1.2.2, 1.2.3, 2.2.3, 2.3.1, 2.3.2, 2.3.3., 3.1.4, 3.2.5,	None	None
Irish Pelagic Sustainability Association	30 October 2008/14 July 2010	– PIs 1.2.1, 1.2.2, 1.2.3, 2.3.1, 2.3.2, 2.3.3, 3.2.2	Mevagissey Fisherman's Association (non-material)	None
Danish Pelagic Producers Organisation	6 November 2008/16 July 2009	– PIs 1.2.2, 3.1.3, 3.2.1, 3.2.2 (all about management plan/HCR)	None	None
Faroese Pelagic Organisation	21 May 2009/12 April 2011	– PIs 1.2.2, 3.1.3, 3.1.4, 3.2.1, 3.2.2***	None****	Marine Scotland/Scottish Government (objection accepted by the Independent Adjudicator)

(continued)

Table 4.3 (continued)

*Fishery/client** *Initial assessment*	*Announced/certified*	*Conditions*	*Stakeholder submissions***	*Objections*
Swedish Pelagic Federation Producer Organisation	6 April 2010/3 September 2011	– PIs 1.2.1, 3.1.1, 3.2.2	None	None
Northern Ireland Pelagic Sustainability Group	4 November 2014/7 November 2016	– PIs 1.2.2, 3.1.1	None	None
Icelandic Sustainable Fisheries	22 December 2016/10 October 2017	– PIs 1.2.2, 2.4.1, 2.4.2, 3.1.3	None	None
Re-assessment				
Mackerel Industry Northern Sustainability Group*****	10 July 2014/10 May 2016	– PIs 1.2.2, 3.1.1	None	None
Faroese Pelagic Organisation******	31 March 2015/15 June 2016	– PIs 1.2.2, 3.1.1	None	None

*The official names of the fisheries in the MSC system and the clients have been shortened here to save space

**Only submissions following the publication of the PCDR are included in the table; many fisheries also received stakeholder submissions at earlier stages of the assessment. Technical oversight comments from MSC are not included although formally they fall under stakeholder comments

***There was an objection to the fishery which the Independent Adjudicator accepted. The conclusion was that the fishery failed. The assessment report is published even if a fishery fails—conditions for other PIs remain, but no condition is set for the PI that fails

****In their objection to this fishery, Marine Scotland noted that they had not been contacted as a potential stakeholder during the assessment, and that they would have submitted comments if they had been

*****Coalition of Danish Pelagic Producers Organisation, Irish Pelagic Sustainability Association, Irish Pelagic Sustainability Group, Norwegian Fishermen's Association, Pelagic Trawler Freezer Association (Netherlands), Scottish Pelagic Sustainability Group and Swedish Pelagic Federation Producer Organisation, i.e. seven of the ten previously certified clients

******Strictly speaking, this was the initial certification of the Faroese fisheries: it had failed its first attempt, following the Marine Scotland objection accepted by the Independent Adjudicator. But it underwent a full assessment at that point also complete, two to three years, and included many conditions, stakeholder submissions and objections. The assessment reports are voluminous: the Ross Sea report is more than 600 pp. long, almost half of which concern the objection process. Stakeholder submissions were also extensive; many were similar to scientific papers. Two additional toothfish fisheries followed with applications in 2009 and 2010, the Syndicat des Armements Réunionnais de Palangriers Congélateurs (SARPC) toothfish fishery, and the Australian Heard Island and McDonald Islands toothfish fishery (with a longliner association and an individual company, respectively, as clients).

The evidence available is overwhelming, and it goes far beyond the space available for this book to give a full overview of the substantive reasons for the conditions, stakeholder submissions and objections.[3] But we see a common thread in all this material, for both the krill and toothfish fisheries: concern about lack of information. Many stakeholders, most of them scientists and NGOs, cite scientific uncertainty about population dynamics and ecosystem relations. Illegal, unreported and unregulated (IUU) fishing has been an issue in some of the toothfish fisheries, likewise the lack of proper tools for species identification. It was also pointed out that the toothfish fishery was categorized by CCAMLR as 'exploratory' rather than 'assessed'. For krill, it was argued that there was insufficient data to ensure that krill or its predators are not harmed by fishing pressures.

All the three first fisheries to achieve MSC certification were subject to objections. At the time of the South Georgia Patagonian toothfish fishery assessment, the MSC had not yet created the position of Independent Adjudicator, so the objection from the National Environmental Trust and The Antarctica Project was handled by an Objection Panel set up by the MSC. The character of the process was more akin to conciliation than adjudication, in practice functioning like another round of peer review, albeit more authoritative. In the event, the assessment report was returned to the CAB (Moody Marine) for reconsideration of some of the scores, but not all those demanded by the objectors. For the Ross Sea assessment (also by Moody Marine), the objection by the Antarctic and Southern Ocean Coalition (ASOC) was handled by the first MSC Independent Adjudicator, Michael Lodge. The objection process took

[3] The improvement conditions are further discussed in Chapter 5.

10 months and involved three separate rounds of elaborate decisions by the Independent Adjudicator. The objector had demanded reductions in the scores of 36 PIs: four of these were accepted; the remainder were dismissed as unfounded. ASOC also objected to Moody Marine's assessment of the Aker Biomarine krill fishery. Unfortunately, the objection documents are not included in the PCR for this fishery, but the conclusion of the Independent Adjudicator is referred: the scores of three PIs had to be revised and a new condition introduced (on the ecosystem effects of the fishery).

All five fisheries succeeded in meeting all conditions set for them during the five-year period during which the MSC certificate is valid, and all applied for re-certification. Now average assessment duration was down to 13 months (median 12 months); three of the fisheries received no conditions, the other two fisheries just one; there was on average one stakeholder submission per fishery, of which two were non-material; and there were no objections.

Cod and Haddock in the Barents Sea

The two first Barents Sea fishery clients to apply for MSC certification, both in 2008, were the Norwegian Seafood Council and the Hong Kong-based Russian company Ocean Trawlers/Three Towns Capital (henceforth: Ocean Trawlers) (see Table 4.2). This difference in client structure reflects the different approaches to MSC certification taken by the two major fishing nations in the Barents Sea. In Norway, industry organizations such as the Norwegian Seafood Council and (later) the Norwegian Fishermen's Association pay for the assessments and let the entire Norwegian fleet be covered by the certificate. In Russia, each individual company seeks its own certificate.

The two initial assessments took 19 and 22 months to complete, respectively. The average (and median) duration of the initial assessments of Barents Sea cod and haddock fisheries was 17 months. The Norwegian fishery had three conditions attached to it, all for P2 (relating to bycatch, ETP species and habitats); the Russian fishery had six, two for each principle. Of the latter, the P1 and P3 principles were quickly closed at the first annual surveillance audits, whereas the P2 conditions, on bycatch and habitats, remained open until the fourth surveillance audit.[4] All conditions for both fisheries were closed during the first certification period, but the Norwegian fishery had two new P2 conditions at re-assessment; the Russian fishery no longer had any conditions. The other Russian fisheries that followed had either no conditions or conditions on P2 only, particularly related to ETP species and habitats.

Norway and Russia conduct the majority of the Barents Sea fisheries. The two coastal states keep approximately 85% of the cod TAC for themselves, but several other states are also involved. After the first Norwegian and Russian fisheries had been certified in 2010, French, Faroese, German, Icelandic and UK fisheries followed suit in 2012, a Spanish fishery in 2013 and a Greenlandic one in 2015. With this, practically all the Barents Sea cod and haddock fisheries were MSC-certified. With the exception of the Greenlandic fishery, none of the fisheries from outside the coastal states received any stakeholder submissions. Apart from the combined UK/German (UK/Deutsche Fischfang-Union (DFFU)/Doggerbank) fishery, which also had some conditions on P3 related to the status of the Fisheries Protection Zone around Svalbard,[5] all conditions concerned P2, in particular PI 2.4 related to habitats.

[4]The P1 conditions concerned the Joint Norwegian–Russian Fisheries Commission departing from the restraint in the HCR not to raise the cod TAC by more than 15% from one year to another, which it had done in a year with exceptionally high stock biomass. The P3 conditions required lobbying from the client in order to propagate precaution in Russian fisheries legislation and the inclusion of NGOs in consultation mechanisms.

[5]International disagreement on the status of the Fisheries Protection Zone around Svalbard has not been an issue in any of the other Barents Sea assessments, as the assessment teams have observed that the legal disagreement among states has not had adverse impacts on the effectiveness of the management of the Barents Sea fisheries. In the UK/DFFU/Doggerbank fishery, the assessment team, noting that it was not competent to make any legal judgment of the issue, leaned on the fishery client's opinion and concluded that the quota distribution in the Svalbard zone was 'unfair' (1-FA, p. 48) (a rationale that would hardly have passed peer review and MSC Technical Oversight scrutiny today). The condition required the client to engage with EU bodies of governance and urge them to bring 'the Svalbard dispute' up with the Norwegian authorities. According to the ensuing annual surveillance reports, the client did that—without the 'dispute' being solved or the EU authorities even making an effort to bring this up with Norway. The condition was not repeated when the fishery was re-certified five years later. In later versions of the MSC Standard, the requirements for harmonization among fisheries have become much stricter; it would no longer be possible for only one fishery to have a condition for an alleged shortcoming that is common to all fisheries in the area.

Of the 12 fisheries that have undergone initial assessment in the Barents Sea cod and haddock fisheries, seven received one or more stakeholder submissions. Six were from regional offices of the World Wide Fund for Nature (WWF): WWF-Germany, WWF-Norway and WWF-Russia. WWF's comments to the first two assessments were extensive and spanned all three MSC Principles; all subsequent submissions, including to the ensuing re-assessments, concerned habitat effects only. That was also the topic of the two objections lodged in the initial assessments (to the Greenlandic fishery and one of the Russian companies, Murmanseld 2)—both by WWF-Germany, the last one jointly with WWF-Russia. The objection to the Greenlandic fishery was lodged in October 2014 and formally dismissed by Independent Adjudicator Michael Lodge five months later, after extensive communication between the parties. During this exchange of proposals and counter-proposals, the CAB (Intertek Fisheries Certification) agreed to reduce some of its scores and introduce a new condition on PI 2.4.3 related to the availability of information on bottom habitats. WWF filed a similar objection to the re-assessment of Ocean Trawlers, which commenced in 2014 and was completed two years later, related to the scoring of PIs 2.4.1 and 2.4.2. A Notice of Objection was submitted in April 2016 but formally withdrawn by the objector five months later. After extensive written consultations between the CAB (Acoura Marine), the client and the objector, facilitated by Independent Adjudicator Eldon V.C. Greenberg, the parties reached an agreement at a physical meeting on the eve of the scheduled oral hearings. No changes were made in the scoring of the fishery, but several clarifications were added to the text of the assessment report.

Three years later, WWF-Germany and WWF-Russia jointly objected on similar grounds to the certification of Russian company Murmanseld 2, then undergoing initial assessment by the certification body DNV GL. Essentially, WWF argued that the assessment team had failed to identify vulnerable marine ecosystems within the area of the fishery and had scored the habitat performance indicator without considering quantitative research data. This objection went through the entire objection procedure laid down in the FCP: initial consultations and exchange of written representations, followed by formal adjudication including oral hearings and subsequent correspondence, before the final decision of the Independent Adjudicator. In the event, Independent Adjudicator John McKendrick QC, basing his decision of the parties' submissions, the MSC Fisheries

Standard and Interpretations Log,[6] as well as 'principles of English [...] administrative law',[7] largely accepted the premises for the WWF objection. In his 25-pp. Post Hearing Decision, he concluded by remanding the case to the CAB for re-scoring and introduction of two new conditions to the certificate. WWF objected to this conclusion as they had wanted a larger reduction in scores, but in his Final Decision the Independent Adjudicator dismissed that objection and accepted the CAB's revised Final (Assessment) Report. As a result of the objection, new conditions were attached to the certificate, but they were not as strict as demanded by WWF.

Mackerel in the Northeast Atlantic

Compared to the Southern Ocean and the Barents Sea fisheries discussed above, a rather different picture appears in the mackerel fishery in the Norwegian Sea (see Table 4.3). A relatively high number of fisheries entered the assessment around 2008. The clients were pelagic fisheries associations from Denmark, the Faroe Islands, Ireland, the Netherlands, Norway and Scotland. Associations from Iceland, Northern Ireland and Sweden followed a few years later. The average (and median) assessment duration was 17 months. Quite a few conditions were attached to these fisheries, related mainly to the challenge of slippage in pelagic fisheries (discard of fish before it enters the vessel) and the implementation of a new harvest control rule set for the mackerel fishery at that time. Notably, none of these fisheries received any material stakeholder submissions,[8] which indicates that no major NGOs had focused on these fisheries, unlike other fisheries in the Northeast Atlantic at the time.[9]

[6] The MSC Interpretations Log (https://mscportal.force.com/interpret/s/) is a tool for CABs, assessment teams and the ASI to ensure consistent interpretation of the Standard and its Guidance. It contains authoritative interpretations of specific requirements based on dilemmas that have emerged in the practical application of the Standard. Interpretations are sometimes included in the Guidance in its subsequent five-year revision.

[7] Murmanseld 2 assessment, 1-FA, p. 289.

[8] The Irish Pelagic Sustainability Alliance fishery received a submission from a local fishermen's association, which was in essence a question about MSC procedures.

[9] This includes the Barents Sea cod and haddock fisheries and the lumpfish fisheries in Greenland, Iceland and Norway, which received substantial attention from environmental NGOs, including several objections. See Hønneland (forthcoming). A summary of the effects of MSC certification on these fisheries is provided below.

In June 2010, an unusual situation arose: the public authorities of a country filed an objection to an MSC assessment of a fishery from another country, on the basis of political disagreement at the state level. (Objectors would normally be NGOs or commercial actors, such as competing companies.) It was Marine Scotland—Scotland's executive agency for fisheries—on behalf of the Scottish Government that objected to the preliminary results (as presented in the PCDR) of the assessment of the Faroese mackerel fishery by the certification body DNV (now DNV GL). As described above, the MSC objection procedure includes several stages where agreement between the parties is encouraged. This objection went through all these stages, including preliminary consultations, full written adjudication and oral hearings—lasting nearly eight months—without an agreement being reached. Hence, the case ended with a final decision by the Independent Adjudicator. Quite unusually, the objection received unsolicited letters of support from industry and science organizations in other countries, albeit with an interest in the fishery themselves (the Norwegian Fishermen's Sales Organization for Pelagic Fish, the Norwegian Shipowners' Association and EU's Pelagic Regional Advisory Council), while the fishery client received formal support from its own public authorities, the Faroese Ministry of Fisheries. Hence, the objection was raised to the political level in Scotland and the Faroe Islands, with semi-public institutions from coastal states Norway and the EU intervening in support of the objector.

In its Notice of Objection, Marine Scotland claimed that the passing score of PI 3.1.1 on the existence of a well-functioning overall management framework was not warranted. The assessment team had referred to the tripartite agreement between the EU, the Faroe Islands and Norway as the basis for its conclusion that this requirement had been met. Marine Scotland, for its part, asserted that as of 1 January 2010 that agreement no longer existed, due to the withdrawal of the Faroe Islands. It also invoked the MSC scope criterion (see the previous section) that a fishery cannot be certified if it operates under a 'controversial unilateral exemption to an international agreement' (FCP 7.4.3)—which it claimed the Faroes withdrawal from the coastal states agreement amounted to.

The fishery client responded that the Faroe Islands had worked decisively to revise the 2000 agreement, which gave EU and Norway 90% of the total allowable catch (TAC), with only 5% each to the Faroe Islands and Russia. The basis for this claim was that far more mackerel had become available in Faroese waters in recent years; moreover, it had become clear that spawning also took place there. As the result of stalled negotiations, the EU and Norway had denied Faroese vessels access to their waters, where Faroes vessels traditionally fish during parts of the year—making it necessary for the Faroe Islands to unilaterally increase their quota share for fishing in their own maritime zones. In its submission to the Independent Adjudicator, the Faroes Ministry of Fisheries claimed that Marine Scotland was incorrect in stating that the Faroe Islands had withdrawn from an international agreement. According to the Ministry, the trilateral cooperation involving the EU, the Faroe Islands and Norway in the management of mackerel had merely been an ad hoc annual arrangement.

The major party opposing the objector during an MSC objection process is not the fishery client, but the CAB. Throughout the objection process, the DNV had consistently rejected the arguments of Marine Scotland. Admittedly, it lowered the score on PI 3.1.1 from 80 (unconditional pass) to 70 (pass with conditions) during this time. However, that was not a result of the objection, but of harmonization discussions with CABs responsible for other mackerel fisheries, due to the deteriorating situation in the coastal states negotiations.

In her Final Decision, Independent Adjudicator Melanie Carter dismissed Marine Scotland's claim that the fishery was out of scope for MSC certification. She did not address the substantive issue of whether the actions of the Faroe Islands represented a controversial unilateral exception to an international agreement, as the scope for certification according to the MSC Standard is determined prior to onset of the assessment—which in this case was before it had become clear that the coastal states would not reach agreement for 2010. However, she concluded that 'the lack of an agreed allocation key for the Total Allowable Catch ("TAC") for 2010 is such that the score for Performance Indicator 3.1.1 cannot be justified on the grounds that it was unreasonable for the Certification Body to have come to this view (p. 244)'. The objection was accepted—and the Faroese mackerel fishery was not certified.

At the time of the Faroese assessment, mackerel catches had already been in excess of the advice of the International Council for the Exploration of the Sea (ICES) for several years. Iceland had in the two preceding years taken for itself a quota of 20% of the TAC recommended by ICES, whereas Norway, the EU and the Faroe Islands had set quotas that did not take the Icelandic quota into consideration. This did not go unnoticed by the CABs responsible for the Northeast Atlantic mackerel fisheries assessments. A harmonization meeting among the CABs was held in February 2010, where it was agreed to invoke a condition on PI 1.2.2, due to fishing mortality being higher than envisaged by the harvest control rule. (This came in time for the condition to be included in the Faroese PCDR.) A condition introduced as the result of extraordinary events during the course of a certification period can be given a timeline of less than the ordinary five years. The new condition to the mackerel fishery was to be met by the end of 2011. That did not occur; on 30 March 2012, all mackerel fisheries in the Northeast Atlantic were suspended from MSC certification.

To allow the continued validity of the certificate, despite its suspension, certified fisheries are required to present and implement a Corrective Action Plan. The NEA Atlantic Mackerel MSC Corrective Action Plan was accepted on 26 June 2012. Seven of the originally certified (now suspended) mackerel clients created the Mackerel Industry Northern Sustainability Alliance (MINSA) to implement the Corrective Action Plan. The Plan contained four elements: lobbying, science, trade-related measures, and media interaction. In brief, the clients agreed to engage with management bodies, scientists and media to promote the need for a coastal states agreement for mackerel. The deadline for implementation of the Corrective Action Plan was 30 April 2014. Following a variation request to the MSC, however, the harmonized two-year condition from 2010 was extended to a regular five-year condition, with a deadline 30 April 2015. On 10 July 2014, the re-assessment of the seven fisheries now part of MINSA was announced (with two UoAs: one for the Norwegian and one for the EU component of the fishery). The deadline for lifting the suspension (in effect, to avoid withdrawal of the certificate) was extended to 10 June 2015, to allow for harmonization discussions with the two mackerel fisheries now undergoing their first full assessment: the Northern Ireland Pelagic Sustainability Group and the Faroese Pelagic Organisation (the latter making a new try after failing its first attempt). On 10 June, however, the CAB (Acoura Marine, now Lloyd's Register) announced that the suspension of all Northeast Atlantic mackerel fisheries was extended until such time as the MINSA fishery should complete its re-assessment: 'At the end of the MINSA assessment, a) if MINSA is certified, new certificates would be issued and the suspension would be lifted; or b) if MINSA is not certified, previous certificates would expire/be withdrawn at that point.'[10] On 10 May 2016, MINSA had passed assessment, without any stakeholder submissions or objections. The Faroese fishery followed a month later.

All the mackerel fisheries were again MSC-certified, four years after their certificates had been suspended—but without the fundamental problem of coastal state agreement having been solved. With fresh five-year conditions on PIs 1.2.2 and 3.1.1, basically requiring the clients to continue lobbying, the deadline for having a coastal states agreement in place was effectively postponed until 2021, ten years after the Independent Adjudicator had ruled that the mackerel fisheries should not be certified without coastal state agreement on allocation keys.

Then in 2019 came a new suspension, but now for a different reason: the declining mackerel stock.

Inshore Arctic Lumpfish in Greenland, Iceland and Norway

Unlike the Southern Ocean and Barents Sea fisheries, which are operated by large trawlers plying the oceans for months on end, the Arctic lumpfish fishery is a typical small-scale, local fishery. It is conducted by small vessels, often one-man boats, using gillnets in fjords and close to the shore. Lumpfish is caught for its roe, which is generally exported. The first Arctic lumpfish fishery to enter MSC assessment was the Icelandic fishery, in 2013 (see Table 4.4). The assessment took 22 months to complete. The Greenlandic fishery followed a year later and the Norwegian in 2016.[11] These took 16 and 15 months, respectively. All fisheries received stakeholder comments (two each for the Greenlandic and Norwegian, and five for the Icelandic), and two of them had an objection lodged against them (the Greenlandic and Icelandic fisheries). Stakeholders providing

[10] Stakeholder announcement—Ref: Extension of the suspension of all NE Atlantic mackerel fisheries within the MINSA group until such time as the MINSA fishery completes re-assessment, Edinburgh: Acoura Marine; available at https://fisheries.msc.org/en/fisheries/minsa-north-east-atlantic-mackerel/@@assessments (accessed 30 June 2020), p. 1. The announcement was signed by representatives of the three CABs involved in the mackerel assessments: Acoura Marine, DNV GL and Intertek Fisheries Certification. It is unclear who made this decision, but the MSC itself was probably involved, and most likely also made the decision, although this does not follow from the announcement.

[11] Individual fishing companies were clients for the Icelandic and Greenlandic fisheries, whereas the Norwegian Fishermen's Association was client for the Norwegian fishery.

Table 4.4 Assessment and re-assessment results of inshore lumpfish fisheries in Greenland, Iceland and Norway certified as per 2020, compiled by the author based on assessment reports available on the MSC website (see reference list)

*Fishery** *Initial assessment*	*Announced/certified*	*Conditions*	*Stakeholder submissions***	*Objections*
Iceland lumpfish	19 February 2013/23 December 2014	– PIs 2.2.2, 2.2.3	– Fuglaverndarfelag Islands (Fuglavernd) (PI 2.2.3) – Environmental Investigation Agency (US) (PIs 2.2.1–2.2.3) – Animal Welfare Institute (US) (PIs 1.2.1, 2.2.2, 2.2.3, 3.1.4) – Royal Greenland and KNAPK (Greenland Fishing and Hunting Association) (procedural) – Dr Reiner Froese, GEOMAR Helmholz-Centre for Ocean Research (general P1 comments)	– Animal Welfare Institute (US) (dismissed on procedural grounds)
Greenland lumpfish	1 April 2014/3 August 2015	– PIs 1.2.1, 1.2.2, 2.2.1, 2.2.2., 2.2.3, 3.2.4	– BirdLife International (PIs 2.2.1–2.2.3) – National Association of Small Boat Owners (general P1/P2 comments)	– BirdLife International (settled by agreement)

*Fishery** *Initial assessment*	*Announced/certified*	*Conditions*	*Stakeholder submissions***	*Objections*
Norway lumpfish	11 July 2016/6 October 2017	– PIs 1.2.2, 1.2.4, 2.3.1, 2.3.2	– BirdLife International (PIs 2.3.1–2.3.3) – WWF-Germany (general comments across all three principles)	None
Re-assessment/second attempt at initial assessment				
Greenland lumpfish	12 August 2019/N/A***	– PIs 2.2.2, 2.2.3, 2.3.2, 3.2.3	N/A***	N/A***
Iceland lumpfish****	18 March 2020/N/A****	– PIs 2.2.1, 2.2.2, 2.2.3, 2.3.2****	N/A****	N/A****

*The official names of the fisheries in the MSC system have been shortened here to save space. Full names are given in the reference list

**Technical oversight comments from the MSC are not included, although formally they fall under stakeholder comments

***The fishery is still under re-assessment at the time of writing (August 2020). The PCDR was published 2 April 2020, but the Final Report has not been published yet

****As the certificate for the Iceland lumpfish fishery was withdrawn 18 April 2019 (having been suspended on 4 January 2018), this is not a re-assessment, but a new attempt at initial assessment. As the new assessment has just been announced at the time of writing (August 2020), scores and indication of conditions are based on the ACDR, hence only provisional

comments ranged from fishing associations, to local and international NGOs and a research institute. The objectors were the US Animal Welfare Institute (for the Icelandic fishery) and BirdLife International (for the Greenlandic fishery).

In its objection to the certification of the Icelandic fishery by the CAB Vottunarstofan Tún, the US Animal Welfare Institute claimed that there was 'significant information lacking in the assessment on known bycatch of cetacean species in this fishery. Entanglements of minke and humpback whales, common dolphins and harbour porpoise in the Icelandic lumpfish gillnet fishery have been noted in both scientific and popular literature'.[12] However, Independent Adjudicator Michael Lodge dismissed the objection on procedural grounds.

BirdLife International's objection during the assessment of the Greenlandic fisheries by DNV GL was also processed quickly, in less than two months. The objection related in its entirety to bycatch of birds, but it addressed all three elements of the bycatch Component of the Assessment Tree: outcome (status of the stocks), management and information. The objector claimed, inter alia, that it is 'near-impossible to quantify bycatch [of, e.g., common eider and common guillemot] adequately and, further, to determine whether this is hindering the full recovery of seabird populations that are increasing in number after over-harvesting'.[13] Also highlighted were the absence of mitigation measures or a strategy to avoid bycatch of birds. Facilitated by Independent Adjudicator Melanie Carter, an agreement was reached between the objector and the CAB, in close communication with the fishery client, after written consultations. As a result, three new conditions were introduced related to bycatch of birds in the fishery: one for the outcome, one for the management and one for the information PI.

The Greenlandic fishery met all its conditions during the five-year certification period and entered re-assessment in 2019. It was re-assessed in 2020 with four new conditions, again all related to the fishery's effects on seabirds.[14] But this time no stakeholder comments were received,

[12] Iceland lumpfish assessment, 1-FA, p. 259.

[13] Greenland lumpfish assessment, 1-FA, p. 261.

[14] One may wonder how new conditions can be introduced at re-assessment, as any problems in the fishery might be expected to be detected at initial assessment and mitigated by the work to close conditions. The situation in the fishery might have changed, but new conditions are generally the result of revisions of the MSC Standard. With every

or objections lodged. The Icelandic fishery, however, never made it to re-assessment. At the third annual surveillance audit, in 2017, new information on seabirds was made available to the assessment team. This was partly the result of data collection initiated by the fishery client (by engaging scientists and independent observers in the fishery) to meet the conditions attached to its MSC certificate. This new information provided evidence to the effect that the 60 score threshold for PIs 2.2.1 and 2.2.2 was no longer met, and the fishery now failed. The certificate was suspended in January 2018 and withdrawn in April 2019. In March 2020, a new attempt at initial certification began.[15] The ACDR, the first draft version of the assessment report published at the time of the fishery's announcement (and the sole draft version available at the time of writing, August 2020), gave only provisional scores; however, at that stage, data supported the assumption that all PIs would achieve a minimum score of 60, and the fishery would pass, with new conditions related to the effects on seabirds and harbour seal.[16]

In summary, the Arctic inshore lumpfish fisheries, just like the Southern Ocean and Barents and Norwegian Seas fisheries, had a considerable number of conditions attached to their first (and some to their second) MSC certificate, some resulting from successful objections by external actors. This chapter has demonstrated how stakeholders such as NGOs and scientists have challenged assessments by CABs and provided new perspectives on the fisheries under assessment. We have also seen that 'complaints', in the form of stakeholder submissions or objections, do not automatically affect the outcome of assessments. The process is evidence-based, and the assessments are conducted by assessors with cutting-edge scientific expertise on specific fisheries. While their scoring of a fishery is not above criticism, it is seldom unfounded. The FCP features several procedural checks and balances, including thorough peer review (in two rounds), Technical Oversight by the MSC and the authoritative decision making of the Independent Adjudicator. Similarly, the opportunities for stakeholder comments ensure external checks and balances, from civil society and the general public.

five-year revision, the requirements of the Standard become stricter—especially for P2, with ecosystem considerations receiving increasing attention.

[15] This time the client was Iceland Sustainable Fisheries (ISF), an industry alliance set up to facilitate MSC certification of all Icelandic fisheries.

[16] Iceland lumpfish assessment, 2-FA, ACDR (draft report), pp. 230–238.

References[17]

AGARBA (Spain) [AGARBA Spain Barents Sea cod] (1-FA 2013, Food Certification International; 1-SA 2015, 2-SA 2016, 3-SA 2017, 4-SA 2018, 1-RA 2019, Bureau Veritas (https://fisheries.msc.org/en/fisheries/agarba-spain-barents-sea-cod/@@view).

Aker Biomarine krill (Norway) [Aker Biomarine Antarctic krill] (1-FA 2010, 1-SA 2011, 2-SA 2012, 3-SA 2013, 4 -SA 2014, Moody Marine; 1-RA 2015, Food Certification International) (https://fisheries.msc.org/en/fisheries/aker-biomarine-antarctic-krill/@@view).

Arkhangelsk Trawl Fleet (Russia) [Arkhangelsk Trawl Fleet Barents and Norwegian Seas cod, haddock, saithe and Greenland halibut] (1-FA 2016, 1-SA 2017, 2-SA 2018, Acoura Marine; 3-SA 2019, Lloyd's Register) (https://fisheries.msc.org/en/fisheries/arkhangelsk-trawl-fleet-norwegian-barents-seas-cod-haddock-saithe-greenland-halibut/@@view).

Danish Pelagic Producers Organisation [Danish Pelagic Producers Organisation North East Atlantic mackerel] (1-FA 2009, 1-SA 2010, 2-SA 2011, 3-SA 2012, 4-SA 2013, DNV GL) (https://fisheries.msc.org/en/fisheries/danish-pelagic-producers-organisation-north-east-atlantic-mackerel/@@view).

Faroe Islands/Iceland [Faroe Islands and Iceland North East Arctic cod and haddock] (1-FA 2012, 1-SA 2013, 2-SA 2014, 3-SA 2015, 4-SA 2016, 1-RA 2017, DNV GL) (https://fisheries.msc.org/en/fisheries/faroe-islands-and-iceland-north-east-arctic-cod-haddock-and-saithe/@@view).

Faroese Pelagic Organisation [Faroese Pelagic Organisation North East Atlantic mackerel] (1-FA 2011, 1-SA 2012, 2-SA 2013, 3-SA 2014, 4-SA 2015, 1-FA [de facto 2nd FA; the first one failed] 2016, DNV GL) (https://fisheries.msc.org/en/fisheries/faroese-pelagic-organisation-north-east-atlantic-mackerel/@@view).

FIUN (Russia) [FIUN Barents & Norwegian Seas cod and haddock] (1-FA 2013, 1-SA 2014, 2-SA 2015, Food Certification International; 3-SA 2016, 4-SA 2017, 1-RA 2018, Acoura Marine) (https://fisheries.msc.org/en/fisheries/fiun-barents-norwegian-seas-cod-and-haddock/@@view).

[17] In this list of MSC assessment reports, the short names given in the tables of this chapter are provided first. Then in square brackets follow the official MSC names, for reference in the MSC website's 'Track a Fishery' function (https://fisheries.msc.org/en/fisheries/). FA = full assessment; SA = (annual) surveillance audit; RA = re-assessment. Hence, 2-SA means the second annual surveillance audit in the five-year certification cycle. The names of the certification bodies, as formal publishers of the reports, are also given. '1-FA 2010, 1-SA 2011, 2-SA 2012, 3-SA 2013, 4 -SA 2014, Moody Marine; 1-RA 2015, Food Certification International' means that all reports up until 2014 were published by Moody Marine, while the 2015 report was published by Food Certification International.

Greenland [Greenland cod, haddock and saithe] (1-FA 2015, Intertek Fisheries Certification; 1-SA 2016, 2-SA 2017, 1-RA 2019, Acoura Marine) (https://fisheries.msc.org/en/fisheries/greenland-cod-haddock-and-saithe-trawl-fishery/@@view).

Greenland lumpfish [Greenland lumpfish] (1-FA 2015, 1-SA 2016, 2-SA 2017, 3-SA 2018, 4-SA 2019, 1-RA 2020, DNV GL) (https://fisheries.msc.org/en/fisheries/greenland-lumpfish/@@assessments).

Heard Island and McDonald Islands toothfish (Australia) [Australian Heard Island and McDonald toothfish and icefish] (1-FA 2012, 1-SA 2013, 2-SA 2014, 3-SA 2015, 4-SA 2016, 1-RA 2017, SCS Global Services) (https://fisheries.msc.org/en/fisheries/australian-heard-island-and-mcdonald-islands-toothfish-icefish-fisheries/@@view).

Iceland lumpfish [Iceland Gillnet lumpfish] (1-FA 2014, 1-SA 2015, 2-SA 2016, 3-SA 2017, Vottunarstofan Tún) (https://fisheries.msc.org/en/fisheries/icelandic-gillnet-lumpfish/@@assessments) Fishery withdrawn in 2017; entered new assessment in 2020 under the name 'ISF Iceland lumpfish' (2-FA 2020, SAI Global) (https://fisheries.msc.org/en/fisheries/isf-iceland-lumpfish/@@assessments).

Irish Pelagic Sustainability Association [Irish Pelagic Sustainability Association (IPSA) western mackerel] (1-FA 2010, 1-SA 2011, 2-SA 2012, 3-SA 2013, 4-SA 2014, Food Certification International) (https://fisheries.msc.org/en/fisheries/irish-pelagic-sustainability-association-ipsa-western-mackerel/@@view).

Irish Pelagic Sustainability Group [Irish Pelagic Sustainability Group (IPSG) western mackerel pelagic trawl] (1-FA 2009, 1-SA 2010, 2-SA 2011, 3-SA 2012, 4-SA 2013, Food Certification International) (https://fisheries.msc.org/en/fisheries/irish-pelagic-sustainability-group-ipsg-western-mackerel-pelagic-trawl/@@view).

ISF Iceland mackerel [Iceland Sustainable Fisheries] (1-FA 2017, 1-SA 2018, 2-SA 2019, SAI Global) (https://fisheries.msc.org/en/fisheries/isf-iceland-mackerel/@@view).

Mackerel Industry Northern Sustainability Group [MINSA North East Atlantic mackerel] (1-FA 2016, 1-SA 2017, 2-SA 2018, Acoura Marine; 3-SA 2019, Lloyd's Register) (https://fisheries.msc.org/en/fisheries/minsa-north-east-atlantic-mackerel/@@view).

Murmanseld 2 (Russia) [Murmanseld 2 Barents Sea cod and haddock] (1-FA 2020, DNV GL) (https://fisheries.msc.org/en/fisheries/murmanseld-2-barents-sea-cod-and-haddock/@@view).

Northern Ireland Pelagic Sustainability Group [Northern Ireland Pelagic Sustainability Group (NIPSG) Irish Sea-Atlantic mackerel & North Sea herring] (1-FA 2016, 1-SA 2017, Acoura Marine; 2-SA 2019, 3-SA 2019, Lloyd's Register) (https://fisheries.msc.org/en/fisheries/northern-ireland-pelagic-sustainability-group-nipsg-irish-sea-atlantic-mackerel-north-sea-herring/@@view).

Norway (Norw. Seafood Council) [Norway North East Arctic cod] (1-FA 2010, 1-SA 2011, 2-SA 2012, 3-SA 2013, 4-SA 2014, 1-RA 2015, DNV GL) (https://fisheries.msc.org/en/fisheries/norway-north-east-arctic-cod/@@assessments).

Norway lumpfish [NFA Norway ling & tusk and NFA Norway lumpfish] (1-FA 2017, Acoura Marine; 1-SA 2019, 2-SA 2020, DNV GL) (https://fisheries.msc.org/en/fisheries/nfa-norway-ling-tusk-and-nfa-norway-lumpfish/@@assessments).

Norwegian Fishermen's Sales Organization for Pelagic Fish [North East Atlantic mackerel pelagic trawl, purse-seine and handline] (1-FA 2009, 1-SA 2010, 2-SA 2011, 3-SA 2012, 4-SA 2013, DNV GL) (https://fisheries.msc.org/en/fisheries/north-east-atlantic-mackerel-pelagic-trawl-purse-seine-and-handline/@@view).

Ocean Trawlers (Russia) [Barents Sea cod, haddock and saithe] (1-FA 2010, 1-SA 2011, 2-SA 2012, 3-SA 2013, 4-SA 2015, Food Certification International; 1-RA 2016, Acoura Marine) (https://fisheries.msc.org/en/fisheries/barents-sea-cod-haddock-and-saithe/@@view).

Oceanprom (Russia) [Oceanprom Barents Sea cod and haddock] (1-FA 2019, DNV GL) (https://fisheries.msc.org/en/fisheries/oceanprom-barents-sea-cod-and-haddock/@@assessments).

Pelagic Freezer Trawler Association (Netherlands) [Pelagic Freezer Trawler Association North East Atlantic mackerel pelagic trawl] (1-FA 2009, 1-SA 2010, 2-SA 2011, 3-SA 2012, 4-SA 2013, Moody Marine) (https://fisheries.msc.org/en/fisheries/pelagic-freezer-trawler-association-north-east-atlantic-mackerel-pelagic-trawl/@@view).

Ross Sea toothfish [Ross Sea toothfish longline] (1-FA 2010, 1-SA 2011, 2-SA 2012, 3-SA 2013, 4-SA 2014, Moody Marine; 1-RA 2015, Acoura Marine) (https://fisheries.msc.org/en/fisheries/ross-sea-toothfish-longline/@@view).

Saint Malo/Euronor (France) [Compagnie des Pêches Saint Malo and Euronor cod and haddock] (1-FA 2012, 1-SA 2013, 2-SA 2014, MacAlister Elliott and Partners Ltd.; 3-SA 2015, 4-SA 2016, 1-RA 2017, ME Certification Ltd.) (https://fisheries.msc.org/en/fisheries/compagnie-des-peches-saint-malo-and-euronor-cod-and-haddock/@@view).

SARPC toothfish [SARPC toothfish (Réunion)] (1-FA 2013, 1-SA 2014, MacAlister Elliot and Partners; 2-SA 2015, 3-SA 2017, 4-SA 2018, ME Certification; 1-RA 2018, Control Union Pesca) (https://fisheries.msc.org/en/fisheries/sarpc-toothfish/@@view).

Scottish Pelagic Sustainability Group [SPSG Ltd western component of north east Atlantic mackerel] (1-FA 2009, 1-SA 2010, 2-SA 2011, 3-SA 2012, 4-SA 2012, Food Certification International) (https://fisheries.msc.org/en/fisheries/spsg-ltd-western-component-of-north-east-atlantic-mackerel/@@view).
South Georgia Patagonian toothfish [South Georgia Patagonian toothfish longline (South Georgia and the South Sandwich Islands/UK)] (1-FA 2004, 1-SA 2005, 2-SA 2006, 3-SA 2007, 4-SA 2008, 1-RA 2009, Moody Marine) (https://fisheries.msc.org/en/fisheries/south-georgia-patagonian-toothfish-longline/@@view).
Strelets/Eridan (Russia) [Russian Federation Barents Sea cod, haddock and saithe] (1-FA 2014, 1-SA 2015, 2-SA 2016, 3-SA 2017, 4-SA 2018, 1-RA 2019, DNV GL) (https://fisheries.msc.org/en/fisheries/russian-federation-barents-sea-cod-haddock-and-saithe/@@view).
Swedish Pelagic Federation Producer Organisation [SPFPO North East Atlantic mackerel] (1-FA 2011, 1-SA 2012, DNV GL) (https://fisheries.msc.org/en/fisheries/spfpo-north-east-atlantic-mackerel/@@view).
UK Fisheries/DFFU/Doggerbank [UK Fisheries Ltd./DFFU/Doggerbank Northeast Arctic cod, haddock and saithe] (1-FA 2012, 1-SA 2013, 2-SA 2014, MacAlister Elliott and Partners Ltd.; 3-SA 2015, 4-SA 2016, 1-RA 2017, ME Certification Ltd.) (https://fisheries.msc.org/en/fisheries/uk-fisheries-ltd-dffu-doggerbank-northeast-arctic-cod-haddock-and-saithe/@@view).

CHAPTER 5

Influence on Target (Fisher) Behaviour and State Practice

Abstract The effects of MSC certification on fisher behaviour have been greater than the effects on state practice. The Southern Ocean fisheries had suffered from lack of information on fish stocks and the ecosystem—MSC certification led to new research initiatives and voluntary collection of data by fishers, and new epistemic communities were formed. However, certification did not lead to any new public regulations. In the Barents Sea, MSC certification directed attention to the fishery's effects on the marine environment, especially bottom habitats. MSC fishery clients in Russia joined forces to develop new fishing gear and voluntarily avoid potential new fishing areas that have not yet been properly researched. This is, however, not reflected in legislation. In the Northeast Atlantic mackerel fisheries, the fishing industry in various European countries joined forces to persuade their respective governments to agree on quotas in line with scientific recommendations, to no avail. The situation is different in the lumpfish fisheries, where states produced new reference points and harvest control rules, and at least some regulations can be attributable to MSC certification.

Keywords Southern Ocean · Barents Sea · Northeast Atlantic · Ecolabelling · Marine Stewardship Council

© The Author(s), under exclusive license to Springer Nature Switzerland AG 2021

G. Hønneland, *Blue Governance in the Arctic and Antarctic*, Palgrave Studies in Maritime Politics and Security, https://doi.org/10.1007/978-3-030-72585-3_5

'A completed condition means a fishery's score meets best practice,' reads the MSC website,[1] accompanied by the information that 92% of certified fisheries have had at least one improvement condition attached to them, to be met during the five-year validity of the certificate. From 2016 to 2018, a total of 288 conditions were set—half of them on P2, the remainder evenly distributed between P1 and P3. The most common type of action required was research, followed by assessment of the fishery's impact and 'technical action' (such as gear modifications).[2]

We now move from structures and processes of MSC certification to the *contents* of the improvement conditions set by the assessment teams, and how fishery clients have gone about meeting the conditions. How has their own behaviour changed, and to what extent have they managed to influence regulatory practice, at the national and international levels? What has been achieved in terms of more sustainable fisheries management and fishing practices? An overview of improvement work to meet conditions in our selected polar fisheries is presented in the first section.[3] We then return to the theoretical discussion initiated in Chapter 2. In the second section, we discuss what form of 'precaution' has been achieved by MSC certification, and in the third section, we examine MSC certification as an implementation mechanism.

Performance Enhancements Following MSC Improvement Conditions

The first round of MSC assessments in the Southern Ocean resulted in a large number of improvement conditions and stakeholder submissions, as well as three extensive objection processes (see Table 4.1). As with the stakeholder submissions and objections, the conditions set by the assessment teams largely concerned the lack of information about the population dynamics of the target stocks and their interaction with the wider ecosystem. Hence, remedial actions have focused on generating

[1] https://www.msc.org/what-we-are-doing/our-collective-impact/fisheries-improving (accessed 5 July 2020).

[2] Ibid.

[3] As in Chapter 4, all information is taken from the MSC assessment reports for the respective fisheries; see reference list for list of reports and their coding for reference. Also here, the material is far too comprehensive to do it justice; we must focus on major trends and examples.

such information, whether through direct measures in fishery clients' own fishing activities (e.g. sampling programmes), support for research (financially or by letting scientists use their vessels) or engagement with the wider epistemic community (including scientists, managers and NGOs). For example, the Aker Biomarine krill fishery had a condition in which the correctness of a set target reference point needed to be validated through further research. The client then organized a working group consisting of scientists from the Norwegian Institute of Marine Research (IMR) and the British Antarctic Survey (BAS), as well as WWF, to address the issue.[4] Another condition called for ascertaining the levels of fish larvae in the krill catch. The client then engaged the consulting firm MRAG, which has special competence on Southern Ocean fisheries, to collect that information through observers onboard the client's vessels—again, IMR and WWF were also invited into the project.[5] More generally, Aker Biomarine started to let IMR use its vessels, free of charge, several weeks outside the fishing season and allowed IMR scientists to join cruises also during the season.[6] It committed to 100% observer coverage, although the CCAMLR requirement was only to have observers on board 50% of the time.[7] Further, Aker Biomarine established a close working relationship with WWF, and was instrumental in the formation of the Association of Responsible Krill Fishing Companies (ARK), which also works to further krill research through cooperation between the industry and scientists.[8] In 2019, Aker Biomarine and other ARK companies contributed to financing the first synoptic survey of Southern Ocean krill since 2000.[9]

[4] Aker Biomarine krill assessment, 1-FA, p. 147.

[5] Ibid., p. 148.

[6] Aker Biomarine krill assessment, 1-RA, p. 11.

[7] Ibid., p. 5.

[8] Ibid., p. 138. See also https://www.rimfrostkrill.com/certifications/association-of-responsible-krill-harvesting-companies-ark (accessed 20 May 2020).

[9] https://www.akerbiomarine.com/blog/krill-industry-antarctic-conservation-in-motion (accessed 5 August 2020). This was not directly related to an MSC condition for the fishery, but the Aker Biomarine first re-assessment report (1-RA, p. 5) highlighted the lack of synoptic surveys since 2000 as the main challenge for the fishery. The more time that passed since the last synoptic survey, the greater the chances that a condition would eventually be introduced to the krill certificates. In financing the survey, therefore, the clients were acting proactively to avoid a future condition.

Similar activities were undertaken in the toothfish fisheries. Among the efforts conducted during the first five-year certification period of the South Georgia fishery was the collection of extensive data on catches of toothfish and bycatch species, improved tools for stock assessment, evaluations of and strategies for the ecosystem impacts of the fishery, and collection of data on benthos.[10] In the Ross Sea fishery, the required actions included assistance to a research project on recruitment dynamics, obligatory participation in a tagging programme for UoA vessels, collection of information on benthic habitats, development of a research plan on trophic effects of the fishery and conducting a risk assessment of the fishery's effects on bycatch populations.[11] One condition required support to the identification of closed areas: here, the client group financed research that contributed to the knowledge basis of what would eventually become CCAMLR's Ross Sea region Marine Protected Area.[12]

Hence, there was extensive engagement by the Southern Ocean fishery clients to meet the conditions set for them, most of which entailed support for research and other information gathering. Admittedly, it is not always clear whether activities described in the surveillance reports were the result of the MSC conditions or whether they would have been conducted otherwise, e.g. at the initiative of CCAMLR or national research institutes. Some actions have clearly been taken by the fishery clients, like contributing finance to the 2019 synoptic krill survey, letting scientific research institutes use their fishing vessels, and voluntarily opting for 100% observer coverage in the krill fishery. This also applies to the sampling programmes and benthos information gathering in the toothfish fisheries. In other instances, client action appears to have come on the back of already planned or ongoing research.[13] Notably, there is no reference in the surveillance audit reports indicating that regulations, whether national or at CCAMLR level, have been amended or established as the result of MSC assessments.

[10] South Georgia Patagonian toothfish assessment, 4-SA, pp. 3–25.

[11] Ross Sea toothfish assessment, 1-FA, pp. 48–50; 4-SA, p. 7.

[12] Ross Sea toothfish assessment, 1-FA, p. 50; 4-SA, pp. 8–9.

[13] There is nothing wrong with that, from the MSC perspective. The point is that improvements are made: where the initiative comes from is less important. However, it matters within the context of this book, where the impacts of private certification schemes are at issue.

In the Barents Sea fisheries, there were a few P1 and P3 condition in the early assessments (see Table 4.2), but most conditions (and all conditions since 2013) have been on P2, in particular related to ETP species and habitats. All stakeholder submissions since 2013, and all three objections, have concerned the impact of bottom trawling on habitats. As noted, half of the clients in the Barents Sea are Russian fishing companies. Remarkably, given their unwillingness to share certificates, all the Russian clients have addressed their habitat-related conditions jointly; the Nikolai M. Knipovich Polar Research Institute of Marine Fisheries and Oceanography (PINRO) and WWF-Murmansk (the local branch of WWF-Russia) are also on board.[14] The cooperation among the companies has involved four elements: first, a new semi-pelagic, 'near-bottom' trawl has been designed and tested, which, if put to use in commercial fisheries, will considerably reduce the impact on bottom habitats.[15] This project is financed by the clients jointly. Second, a programme for the registration of benthic encounters has been designed by WWF and implemented on the client vessels.[16] Third, in 2016, the Russian clients jointly created the Coordination Council for the Development of Sustainable Fisheries in the North and signed an agreement on efforts to minimize impacts of bottom trawl on the benthic ecosystems in the Barents Sea.[17] Fourth, under the Coordination Council agreement, the client companies have committed not to enter new fishing grounds in the Barents Sea (as in the northern parts of the Barents Sea, now becoming more accessible due to changes in ice coverage) until these areas have been appropriately researched.[18] From the information in the surveillance audit reports, there is little doubt that these developments follow directly from the clients' endeavours to meet their MSC conditions.[19] Again, however, there is no evidence that management practices, national legislation or international agreements between Norway and Russia in their Joint Commission have

[14] See, e.g., the FIUN assessment, 1-RA, p. 179.

[15] See, e.g., the FIUN assessment, 4-SA, p. 31.

[16] See, e.g., the FIUN assessment, 1-RA, p. 189.

[17] See, e.g., the FIUN assessment, 1-SA, p. 16.

[18] See, e.g., the FIUN assessment, 4-SA, p. 32.

[19] There is no mention that the relevant actions are required by national legislation or international instruments, and to the best of my knowledge there is no such requirement applicable to the Barents Sea.

been affected by the MSC assessments. All the Norwegian fleet and more than 85% of the Russian fleet are MSC-certified[20]; the relevant authorities on both sides are consulted at each annual surveillance audit, so they are well aware that their national fleets are in the MSC programme. But there is total silence from the authorities on both sides regarding private certification schemes. A search for 'MSC' or 'Marine Stewardship Council' on the websites of Norwegian and Russian fisheries management bodies[21] and in the protocols of the Joint Norwegian–Russian Fisheries Commission[22] yields not a single hit. Private certification is apparently not something Norwegian and Russian authorities wish to flag as part of the overarching management of their fisheries. Probing the reasons is beyond the scope of this book, but perhaps the MSC is seen as a 'competitor' and/or (foreign) 'intruder' in national fisheries management.[23]

The conditions in the third-state (non-coastal state) fisheries also generally relate to P2, in particular to the protection of ETP species and habitat structures. The conditions to the certificates, which were all met during the five-year certification period, included recording and registration of bycatch beyond what is required by Norwegian and Russian regulations; voluntary adherence to established fishing grounds (documented to the assessment team at annual surveillance audits by satellite-based vessel monitoring system (VMS) logs); production of company-level strategies, codes of conduct and action plans to avoid encounters with sponge and coral garden habitats; and obligatory training of crews on identifying ETP species and habitat structures.[24]

In the small-scale lumpfish fisheries, the direct effects of MSC certification are even more immediately apparent. Problems identified at initial

[20] All Norwegian vessels are covered by the Norway North East Arctic cod certificate. The number of Russian vessels certified is approximate, based on numbers of vessels indicated in the various assessment reports.

[21] https://www.regjeringen.no/no/dep/nfd/id709/ (Ministry of Trade, Industry and Fisheries, Norway), https://www.fiskeridir.no/ (Directorate of Fisheries, Norway), http://www.fish.gov.ru/ (Federal Fisheries Agency, Russian Federation) (all accessed 20 May 2020).

[22] https://www.jointfish.com/ (website of the Joint Norwegian–Russian Fisheries Commission) (accessed 20 May 2020).

[23] See discussion in Chapter 6 on the 'distribution of work' between governmental action and private certification.

[24] See, e.g., Greenland cod and haddock assessment, 3-SA, pp. 18–19.

certification were lack of appropriate reference points, HCRs and management plans, as well as strategies for reducing bycatch of seabirds and marine mammals (see Table 4.4). New reference points and HCRs were produced in Greenland and are underway in Norway.[25] Monitoring of bycatch of seabirds (in Iceland, also of marine mammals) has increased in all three countries, and new mitigation strategies have been produced.[26] Anecdotal evidence suggests that MSC certification is a 'to be or not to be' requirement to get lumpfish roe sold on the global market, so fishing companies go to great lengths to get the authorities to adopt the precautionary measures necessary for achieving and retaining MSC certification.[27] An April 2020 press release from the Icelandic Ministry of Fisheries and Agriculture was explicit about the role of certification in connection with a precautionary seasonal halt of the lumpfish fishery: this was said to be due to the need to keep fishing 'in accordance with scientific advice' and 'to ensure that the existing certifications are not lost'.[28] Here we can note a direct link between MSC requirements and national regulations, as clearly stated by the relevant authorities themselves.

In the Northeast Atlantic, ten mackerel clients from nine countries (two from Ireland) have undergone MSC assessment. In the first round of assessments, slippage and implementation of a new management plan were identified as areas for improvement (see Table 4.3). Clients were required to report slippage and produce strategies for reducing it, which they did. Overshoot of the recommended TAC was an emerging issue also before the Faroe Islands (in the view of the other states) withdrew from the established regime by unilaterally setting its national quota. So all MSC fishery clients received a condition that required them to 'work with' their respective national authorities to persuade these states to come to agreement. A typical formulation from an annual surveillance

[25] Greenland lumpfish assessment, 4-SA, pp. 26–27; Norway lumpfish assessment, 2-SA, pp. 37–43.

[26] Greenland lumpfish assessment, 4-SA, pp. 28–30; Iceland lumpfish assessment, 3-SA, pp. 17–20; Norway lumpfish assessment, 2-SA, pp. 39–43, p. 51.

[27] Anecdotal evidence, based on the author's own experience. (See Chapter 1 on methodology.).

[28] https://www.stjornarradid.is/efst-a-baugi/frettir/stok-frett/2020/04/30/Breyting-a-reglugerd-um-veidar-a-grasleppu-2020/ (accessed 30 April 2020). The English formulation follows from Google Translate. The statement was imprecise: the certificate was already lost, but assessment was underway to get it reinstated.

audit report on a client's progress towards meeting their conditions is the following:

> SPSG has vigorously canvassed for settlement of a coastal states agreement for the division of mackerel quota within the agreed TAC, but to date such an agreement has yet be achieved. The SPSG has been an active participant in twelve negotiating sessions intended to bring about a settlement of quota allocations between the relevant coastal states. (SFSG fishery assessment, 4-SA, p. 15)

From 2010 onwards, the fisheries received a new condition under P3 (either PI 3.1.1 or 3.2.2), stating that a new coastal states agreement, bringing total catches in line with ICES advice, must be in place within five years for the fisheries to remain certified. As noted, the mackerel certificates were suspended in 2012 but were reinstated when the 'all-European' MINSA fishery was certified in 2016. A coastal states agreement was not any nearer than in the preceding years, but the conditions—on PI 1.2.2 related to the management plan and PI 3.1.1 on the overarching management framework—were now reset to point zero: an agreement would have to be in place within five years. In the meantime, the clients were obliged to continue 'working with the authorities'—the surveillance audit reports are full of references to the many meetings attended by the clients and the letters sent to ministries, directorates, commissions and others, urging governments to come to an agreement. It was all to no avail; as of autumn 2020, a coastal states agreement is still not in sight. As noted, the Northeast Atlantic mackerel fisheries had their certificates suspended again in 2019, this time because the stock size had decreased below acceptable levels.

Of these four cases, mackerel is the one where it is most difficult to see any improvements as a result of MSC certification. Little has been achieved, beyond updated information on slippage in the fishery and company strategies on how to reduce it. Of course, it might be argued that it is an improvement in itself that the fishing industries in eight European countries became driving forces in arguing for sustainable TAC levels. Fisher behaviour has indeed been affected—but in the corridors of negotiations, not in the fishing fields. However, decade-long endeavours to influence the behaviour of decision-makers have proven futile.

In the mackerel fisheries, the MSC itself has been more active than in other Arctic fisheries assessments. Normally, the MSC limits its role to producing the certification requirements, providing technical oversight comments to assessment reports, and issuing certificates. In the Northeast Atlantic mackerel fisheries, however, the MSC apparently involved itself in the harmonization of scores between fisheries (usually done by the CABs alone) and, not least, the prolongation of certificates beyond what is usual. After four years of suspension, the mackerel certificates were reinstated in 2016, despite no progress in the coastal states negotiations, where the stalemate was the reason for suspension in the first place. The deadline for having a coastal states agreement in place was effectively postponed until 2021, ten years after the Independent Adjudicator had ruled against certification of the Northeast Atlantic mackerel fishery due to the lack of a coastal states agreement. Meanwhile, the fisheries remained certified—until the stock itself fell below acceptable levels and the fishery was suspended for that reason in 2019.

The mackerel fishery might have become one of MSC's big success stories, if the engagement of all the clients with their national authorities had produced the intended effects. This would have been the story of how the fishing industry across northern Europe joined forces to convince their respective national authorities to come to an agreement, which would have been a massive demonstration of the power of private certification. As that failed to happen, the mackerel story is perhaps not the MSC's finest hour—what remains is the image of a scheme owner that may have gone a bit too far in seeking to keep one of the biggest, classic European fisheries within its portfolio.

Precautionary Outcomes of MSC Certification

As seen in Chapter 2, a common thread in the literature on the precautionary approach related to both international environmental law and international fisheries law is the division into 'strong' or 'narrow' and 'weak' or 'broad' versions of the principle.[29] The 'strong/narrow' version is normally seen in relation to the reversal of the burden of proof. In international environmental politics, this version, in its purest sense, allows states to conduct industrial activities only insofar as they can prove that damage is not caused to the environment (beyond a set limit); in fisheries management, it implies that the authorities can allow fishing only

[29] See, e.g., Bodansky (2004), Ebben (2011), Gillespie (2007) and Sands, and Peel (2018). For a further discussion, see Chapter 2.

when they can document that the fishery does not affect the target stocks or the marine environment (beyond pre-agreed boundaries). By contrast, the weakest/broadest versions of the precautionary approach prescribe due diligence or prudent foresight of some sort, without defining any quantitative benchmarks for action.

As also seen in Chapter 2, the FSA is rather specific in its provisions concerning the precautionary approach. The biological reference points are a cornerstone in the FSA's operationalization of the approach. Under Article 6(3)(b), states are obliged to determine stock-specific (target and limit) reference points (usually for the size of the spawning stock and for fishing mortality) and the action to be taken if these are exceeded. The latter, in practice, entails producing an HCR, a harvest control rule. While no further operationalization of the precautionary approach is offered by the CCRF, its Technical Guidelines No. 2 expand on the concept,[30] dismissing the reversal of the burden of proof and claiming that a precautionary approach to fisheries management entails appropriate access control rules, data reporting requirements, planning processes, monitoring and enforcement arrangements, mechanisms for re-evaluation of the management system and implementation guidelines. Likewise, scholars have listed stock assessment, transparency, stakeholder participation, effective monitoring and strict penalties as characteristics of a management system based on the precautionary.[31] Hence, in legally binding instruments (the FSA) and non-legally binding ones (the CCRF), as well as in mainstream academic literature,[32] two versions of the precautionary approach emerge: one 'semi-strong and narrow'[33] and one 'semi-weak and broad'.[34] In brief, the 'semi-strong/narrow'

[30] Precautionary Approach to Capture Fisheries and Species Introductions, FAO Technical Guidelines for Responsible Fisheries, No. 2, FAO, Rome, 1996.

[31] See, e.g., Hewison (1996) and Garcia (1994). For a further discussion, see Chapter 2.

[32] As noted in Chapter 2, there is also a more critical social science literature on the precautionary approach (e.g., Charles 2002; Gerrodette et al. 2002) prescribing a more systematic review of uncertainties in fisheries management, including uncertainties in stock modelling and calculation of reference points.

[33] The version is 'semi-strong' and 'narrow' because it requires reference points, which is the most concrete manifestation of the precautionary approach. However, it does not reverse the burden of proof and hence does not qualify as 'strong'.

[34] By 'semi-weak' and 'broad' is here meant that this version is not as 'strong' as to specifically require reference points, but not as 'weak' as to only prescribe general due

approach commands reference points and HCRs, whereas the 'semi-weak/broad' version merely envisages typical components of a fisheries management system, some of which are traditional (such as stock assessment, access control and enforcement), others of more recent date (such as transparency in decision-making, and stakeholder participation).[35]

The MSC Fisheries Standard offers all this: first, with regard to the 'semi-strong/narrow' approach, stock-specific reference points and HCRs are part and parcel of P1. A fishery may obtain certification with a 'generally understood' (i.e. implicit) HCR in place, based on target and limit reference points, but not without one. And it will pass without condition only with a 'well-defined' (i.e. explicit) HCR. Hence, if there is an implicit HCR in place, an explicit HCR must be produced during the first five-year certification period or the certificate will be suspended. Second, with regard to the 'semi-weak/broad' precautionary approach, the MSC Fisheries Standard ticks all the boxes with its requirements regarding, inter alia, stock assessments (PIs 1.2.3 and 1.2.4), protection of the wider marine ecosystem (all of P2), stakeholder engagement (PI 3.1.2), enforcement and compliance (PI 3.2.3) and review of the management system (PI 3.2.4)—and formal application of the precautionary approach itself (PIs 3.1.3 and 3.2.2).

What, then, has been achieved in the polar fisheries under assessment here? From the perspective of the 'semi-strong/narrow' approach, MSC certification has (i) been an attestation to the appropriateness and soundness of the established reference points and HCRs in the Barents Sea fisheries; (ii) helped to ascertain the appropriateness and soundness of HCRs in the Southern Ocean fisheries; and (iii) been the direct cause of the development and establishment of reference points and HCRs in the Arctic lumpfish fisheries. Thus, MSC certification has not been without effect in this regard, with the greatest impact (not unexpectedly) in fisheries where reference points and HCRs were not previously in place. As far as the 'semi-weak/broad' approach is concerned, several improvements have been made to management systems as a result of conditions to MSC certification. In addition to the reference points and HCR-related issues, a substantial amount of new research and information gathering has been

diligence. Instead, it demands typical components of a fisheries management system, such as stock assessment, access control and enforcement; see discussion above.

[35] See, e.g., FSA, Art. 7 on the requirement to transparency in RFMOs.

conducted in the Southern Ocean, and efforts have been made to reduce impacts on bottom habitats in the Barents Sea and on seabirds and marine mammals in the Arctic lumpfish fisheries. There have been very few, if any, conditions related to the status or management of, or information on, the target or bycatch stocks (Components 1.1, 2.1 and 2.2), the overarching layout of the management system (with the exception of the mackerel fisheries), possibilities for stakeholder engagement (PI 3.1.2), the application of the precautionary approach (PI 3.1.3), the appropriateness of fishery-specific objectives and decision-making procedures (PIs 3.2.1 and 3.2.2), the effectiveness of the enforcement system and level of compliance (PI 3.2.3), or the existence of internal and external reviews of the management system (PI 3.2.4). Perhaps there were missed opportunities for improvement here—but it might also simply be that the fisheries perform sufficiently well on these indicators and no MSC 'intervention' is needed. Notably, conditions for *outcome* indicators—i.e. indicators reflecting the *status* of the fish stocks, whether target or bycatch stocks—were practically absent (again with the exception of the mackerel fisheries).[36]

Most fundamentally, the MSC Certification Scheme represents a massive *operationalization* of the precautionary approach. First, it offers a global benchmark for the establishment of reference points and HCRs (cf. the 'semi-strong/narrow' version of the precautionary approach) where previously there was only piecemeal operationalization at national and RFMO levels. Following the adoption of the FSA in 1995, national authorities and RFMOs around the world have worked to implement reference points and HCRs, but the endeavours have not been coordinated or methodologies harmonized. Although still in relatively broad categories, the MSC Fisheries Standard offers just that: a global benchmark for establishing reference points and HCRs. Second, and arguably its most innovative contribution (cf. the 'semi-weak/broad' version), the MSC Fisheries Standard operationalizes the precautionary approach not only for the management of target stocks (P1), which has traditionally been the priority of RFMOs (with CCAMLR as a notable exception), but also for the protection of the marine environment (P2) and for the organization of the management system (P3).

[36]This might indicate that clients seek certification only when they feel certain that the fish stocks are in good shape, while they pay less attention to the soundness of management measures and ecosystem impacts.

An Iterative and Evolving Implementation Mechanism

Traditional implementation of international obligations is the remit of states, individually or in collaboration. Legislation and policies (*output*) are produced, intended to change the behaviour of target groups (*outcome*), in order to achieve the final objectives of the provisions (*impact*). In the fisheries sector, states or (in internationally managed fisheries) RFMOs determine policies, and the national authorities (and sometimes RFMOs) are vested with the authority to enforce regulations. Successful implementation is dependent on states' readiness to adapt their regulations in line with their international obligations, and target groups' readiness to adapt their behaviour in line with these regulations. 'Readiness' here implies both willingness and ability. Sometimes states (or target groups) have no intention of honouring their international obligations; in other cases, they intend to, but lack the resources (financial or practical). That is why international agreements, especially relatively recent ones, sometimes contain compliance mechanisms to assist states in implementing their obligations.[37] Typically, that involves financial assistance, capacity building or coordinated (or joint) implementation.[38] The question here is whether a private certification scheme like the MSC can serve the same purpose—and if so, which mechanisms are at work. What distinguishes private certification from state-driven implementation?

First, there is the obvious fact that the initiatives come from elsewhere and the incentives are different. States are largely sidelined here. There is an international non-profit organization, the MSC, with considerable financial resources and several hundred employees who work to get market actors around the world commit to purchasing only MSC-certified fish. There are the fishers themselves, who respond by getting certified, whether enthusiastically or not. States are, in MSC assessments, defined as stakeholders, alongside, for instance, scientists and NGOs. They can express their views during the assessment process if they wish, and are subsequently courted by fishery clients who need management policies

[37] See Part I of Victor et al. (1998) for examples.

[38] Hønneland (2013, pp. 26–27). At the individual level, states would provide guidance to target groups on how to comply with regulation, in addition to traditional coercive compliance mechanisms (deterrence) (ibid., p. 18).

changed in order for them to remain certified. In brief, states are responsive rather than at the wheel—and they have no obligation, nor much apparent intent,[39] to respond either.

Fishery clients, on the other hand, have everything to gain, or lose, from certification. One of the most remarkable outcomes in this case study, beyond the achievement of substantive precautionary management measures, is the development of *new epistemic communities* in all three polar fisheries studied here, with the fishery clients as the central communication hubs. An epistemic community is 'a network of professionals with recognized expertise and competence in a particular domain and an authoritative claim to policy relevant knowledge within that domain or issue-area'.[40] In international environmental politics, or fisheries management, an epistemic community will typically consist of scientists, government officials, industry representatives and NGOs with a shared set of normative beliefs, assumptions about causality and scientific validity, and a common political enterprise.[41] In the internationally managed Northeast Atlantic fisheries, for instance, epistemic communities may consist of representatives of the International Council for the Exploration of the Sea (ICES), national research institutes, the North-East Atlantic Fisheries Commission (NEAFC), national fishery managers and sometimes also the fishing industry and NGOs. Driving forces will typically be national governments, with the ICES and NEAFC secretariats serving as communication hubs (as the CCAMLR secretariat in the Southern Ocean fisheries).

In all our four fisheries, albeit to varying extents, alternative epistemic communities have developed as a result of MSC certification. The actors are largely the same as in the established networks, but the dynamics are different. In the Southern Ocean krill fisheries, the fishery clients have developed the ARK, which coordinates non-state efforts to collect more information on the krill population, in cooperation with CCAMLR. They were significant contributors to the 2019 synoptic survey of Antarctic krill, the first one in nearly 20 years—this was done in close cooperation

[39] Of the 40-odd stakeholder submissions received for the fisheries studied here (see Chapter 4), none were from national authorities or inter-governmental institutions (except one in the mackerel fisheries [see Table 4.3] and one non-material in the Barents Sea cod and haddock fisheries [see Table 4.2]).

[40] Haas (1992, p. 3).

[41] Ibid., p. 3.

with scientific institutions and national authorities in different CCAMLR states. At the national level in Norway, Aker Biomarine has assumed an informal coordinating role between the industry, national authorities, science and NGOs. As explained in the first section of this chapter, Aker supports both science and NGOs, and organizes meetings and seminars for all the above-mentioned actors, aimed at increasing knowledge about the krill population. The authorities are at the helm of the Norwegian delegation in CCAMLR negotiations, but they acknowledge the role of Aker Biomarine as constructive and significant.[42] In the Barents Sea, the MSC client Ocean Trawlers (now: Norebo) has assumed a leading role in the Russian Coordination Council for the Development of Sustainable Fisheries in the North, which was established to meet the requirements of an MSC condition on habitat impacts of the trawl fishery. All the Russian MSC clients are involved in this initiative, as well as PINRO and WWF. However, the link to the Russian authorities, and hence Russian regulations and policies, is less clear. In the Northeast Atlantic mackerel fisheries, fishing companies in different European countries joint forces to persuade their respective governments to come to agreement on TACs in line with scientific recommendations from ICES. As regards the final cluster of fisheries, the Arctic lumpfish fisheries, Iceland and Norway have national certification units that work to get all national fisheries MSC-certified (not only lumpfish, but *all* species). Iceland Sustainable Fisheries (ISF) was set up with this specific objective,[43] while the Norwegian Fishermen's Association and the Norwegian Seafood Council have jointly facilitated (and financed) MSC certification of all Norwegian fisheries. In both ISF and the Norwegian Fishermen's Association, staff are working full time to organize MSC certifications and follow up existing certificates through annual surveillance audits and re-certifications. An important part of this work is to get national scientific institutions and authorities to contribute

[42] As noted in the section on methodology in Chapter 1, formal interviews are not part of this investigation, but in an informal conversation with the author (name, affiliation, place and date omitted to preserve anonymity), a Norwegian governmental representative said something along the lines of 'We're like a big family [all Norwegian delegation members and observers to CCAMLR] – we respect that each of us has his or her particular role to play, but we all get along well and work towards the same goal'. The proactive and constructive role of the MSC fishery client was particularly emphasized by my conversation partner.

[43] While an individual fishing company was originally client for the Icelandic fishery, ISF is client for the on-going (August 2020) second attempt at certification.

to fulfilment of their MSC conditions, for instance by increasing scientific effort and amending national regulations or policies. Whereas initiatives for stricter regulations once came solely from the authorities, with the fishing industry often fighting against them, the situation is now turned on its head—at least to some extent.

Similarly, MSC certification has affected relationships between various bodies of governance within states. Protection of bottom habitats has been the main issue of concern in the assessments of the Barents Sea fisheries. That has never been a priority area for the Joint Norwegian–Russian Fisheries Commission, nor in the national management systems for fisheries management in the two coastal states—arguably because the marine environment has traditionally been the remit of another branch of government: the environmental authorities. In that regard, MSC certification not only fills a regulatory gap, but may also help to bridge the institutional gap between different sectors of government in the coastal states, by forcing fisheries management authorities to work with environmental authorities to address concerns related to the marine environment.

An extraordinarily important aspect of the MSC Certification Scheme as an implementation mechanism is its *iterative* and *evolving* nature. If certification had been a one-off event—as presumably many people think ecolabelling is—it would have served to distinguish the well-managed fisheries from the less well-managed ones, but without spurring enhanced performance beyond whatever preparations the clients might have made prior to certification. First, fisheries are not MSC-certified once and for all; every five years they must start from scratch with (a new round of) re-certification. Here I have emphasised the improvement conditions that bring fisheries forward, but it should be recalled that fisheries are scrutinized throughout the five-year certification periods, also *between* the annual surveillance audits. In case of developments that might change the scores (e.g. updated stock assessments indicating unforeseen shifts in stock size or dynamics), CABs are obliged not only to address this during the following annual surveillance audit, but to initiate 'expedited audits' without delay to address the changes and, if necessary, re-score the fishery. A fishery can never be complacent about its certification status: if the situation in the fishery deteriorates, a new condition might be introduced at short notice (if the score on a performance indicator falls below 80), or the certificate might be suspended (if the score falls below 60). From a sustainability perspective, this safeguards against an unsustainable fishery remaining certified until its next regular surveillance audit, which might be almost a year ahead.

Second, the MSC Fisheries Standard is an evolving standard. It is revised every five years, and with each iteration the requirements for certification get stricter. Even if a fishery passes certification under one version of the Standard, it might have to make further efforts in order to remain certified at its next re-certification. As shown in Chapter 4, fisheries that had successfully fulfilled their conditions during one certification period were met with new conditions at their next re-certification. Hence, the precautionary approach is progressively operationalized and implemented as the MSC Fisheries Certification Scheme evolves.

References

Bodansky, Dan, 'Deconstructing the Precautionary Principle', in David D. Caron and Harry N. Scheiber (eds), *Bringing New Law to Ocean Waters*, Leiden: Martinus Nijhoff, 2004.

Charles, Anthony T., 'The Precautionary Approach and "Burden of Proof" Challenges in Fisheries Management', 70 *Bulletin of Marine Sciences* 2002, pp. 683–694.

Ebben, Thomas, 'The Implementation of the Precautionary Approach in International Fishery Law: A Move towards "Green" Fishery', 15 *New Zealand Journal of Environmental Law* 2011, pp. 113–146.

Garcia, Serge Michel, 'The Precautionary Principle: Its Implications in Capture Fisheries Management', 22 *Ocean & Coastal Management* 1994, pp. 99–125.

Gerrodette, Tim, Paul K. Dayton, Seth Macinko and Michael J. Fogarty, 'Precautionary Management of Marine Fisheries: Moving beyond Burden of Proof', 70 *Bulletin of Marine Sciences* 2002, pp. 657–668.

Gillespie, Alexander, 'The Precautionary Principle in the 21st Century: A Case Study of Noise Pollution', 22 *International Journal of Marine and Coastal Law* 2007, pp. 61–87.

Haas, Peter M., 'Introduction: Epistemic Communities and International Policy Coordination', 46 *International Organization* 1992, pp. 1–36.

Hewison, Grant J., 'The Precautionary Approach to Fisheries Management: An Environmental Perspective', 11 *International Journal of Marine & Coastal Law* 1996, pp. 301–332.

Hønneland, Geir, *Making Fishery Agreements Work: Post-Agreement Bargaining in the Barents Sea*, Cheltenham: Edward Elgar, 2013.

Sands, Philippe and Jaqueline Peel with Adriana Febra and Ruth MacKenzie, *Principles of International Environmental Law*, fourth edition, Cambridge: Cambridge University Press, 2018.

Victor, David G., Kal Raustiala and Eugene B. Skolnikoff (eds), *The Implementation and Effectiveness of International Environmental Commitments: Theory and Practice*, Cambridge, MA: MIT Press, 1998.

MSC Assessment Reports[44]

AGARBA (Spain) [AGARBA Spain Barents Sea cod] (1-FA 2013, Food Certification International; 1-SA 2015, 2-SA 2016, 3-SA 2017, 4-SA 2018, 1-RA 2019, Bureau Veritas) (https://fisheries.msc.org/en/fisheries/agarba-spain-barents-sea-cod/@@view).

Aker Biomarine krill (Norway) [Aker Biomarine Antarctic krill] (1-FA 2010, 1-SA 2011, 2-SA 2012, 3-SA 2013, 4 -SA 2014, Moody Marine; 1-RA 2015, Food Certification International) (https://fisheries.msc.org/en/fisheries/aker-biomarine-antarctic-krill/@@view).

Arkhangelsk Trawl Fleet (Russia) [Arkhangelsk Trawl Fleet Barents and Norwegian Seas cod, haddock, saithe and Greenland halibut] (1-FA 2016, 1-SA 2017, 2-SA 2018, Acoura Marine; 3-SA 2019, Lloyd's Register) (https://fisheries.msc.org/en/fisheries/arkhangelsk-trawl-fleet-norwegian-barents-seas-cod-haddock-saithe-greenland-halibut/@@view).

Danish Pelagic Producers Organisation [Danish Pelagic Producers Organisation North East Atlantic mackerel] (1-FA 2009, 1-SA 2010, 2-SA 2011, 3-SA 2012, 4-SA 2013, DNV GL) (https://fisheries.msc.org/en/fisheries/danish-pelagic-producers-organisation-north-east-atlantic-mackerel/@@view).

Faroe Islands/Iceland [Faroe Islands and Iceland North East Arctic cod and haddock] (1-FA 2012, 1-SA 2013, 2-SA 2014, 3-SA 2015, 4-SA 2016, 1-RA 2017, DNV GL) (https://fisheries.msc.org/en/fisheries/faroe-islands-and-iceland-north-east-arctic-cod-haddock-and-saithe/@@view).

Faroese Pelagic Organisation [Faroese Pelagic Organisation North East Atlantic mackerel] (1-FA 2011, 1-SA 2012, 2-SA 2013, 3-SA 2014, 4-SA 2015, 1-FA [de facto 2nd FA; the first one failed] 2016, DNV GL) (https://fisheries.msc.org/en/fisheries/faroese-pelagic-organisation-north-east-atlantic-mackerel/@@view).

[44] In this list of MSC assessment reports, the short names given in the tables in Chapter 1 are provided first. Then in square brackets follow the official MSC names, for reference in the MSC website's 'Track a Fishery' function (https://fisheries.msc.org/en/fisheries/). FA = full assessment; SA = (annual) surveillance audit; RA = re-assessment. Hence, 2-SA means the second annual surveillance audit in the five-year certification cycle. The names of the certification bodies, as formal publishers of the reports, are also given. '1-FA 2010, 1-SA 2011, 2-SA 2012, 3-SA 2013, 4 -SA 2014, Moody Marine; 1-RA 2015, Food Certification International' means that all reports up until 2014 were published by Moody Marine, while the 2015 report was published by Food Certification International.

FIUN (Russia) [FIUN Barents & Norwegian Seas cod and haddock] (1-FA 2013, 1-SA 2014, 2-SA 2015, Food Certification International; 3-SA 2016, 4-SA 2017, 1-RA 2018, Acoura Marine) (https://fisheries.msc.org/en/fisheries/fiun-barents-norwegian-seas-cod-and-haddock/@@view).

Greenland [Greenland cod, haddock and saithe] (1-FA 2015, Intertek Fisheries Certification; 1-SA 2016, 2-SA 2017, 1-RA 2019, Acoura Marine) (https://fisheries.msc.org/en/fisheries/greenland-cod-haddock-and-saithe-trawl-fishery/@@view).

Greenland lumpfish [Greenland lumpfish] (1-FA 2015, 1-SA 2016, 2-SA 2017, 3-SA 2018, 4-SA 2019, 1-RA 2020, DNV GL) (https://fisheries.msc.org/en/fisheries/greenland-lumpfish/@@assessments).

Heard Island and McDonald Islands toothfish (Australia) [Australian Heard Island and McDonald toothfish and icefish] (1-FA 2012, 1-SA 2013, 2-SA 2014, 3-SA 2015, 4-SA 2016, 1-RA 2017, SCS Global Services) (https://fisheries.msc.org/en/fisheries/australian-heard-island-and-mcdonald-islands-toothfish-icefish-fisheries/@@view).

Iceland lumpfish [Iceland Gillnet lumpfish] (1-FA 2014, 1-SA 2015, 2-SA 2016, 3-SA 2017, Vottunarstofan Tún) (https://fisheries.msc.org/en/fisheries/icelandic-gillnet-lumpfish/@@assessments). Fishery withdrawn in 2017; entered new assessment in 2020 under the name 'ISF Iceland lumpfish' (2-FA 2020, SAI Global) (https://fisheries.msc.org/en/fisheries/isf-iceland-lumpfish/@@assessments).

Irish Pelagic Sustainability Association [Irish Pelagic Sustainability Association (IPSA) western mackerel] (1-FA 2010, 1-SA 2011, 2-SA 2012, 3-SA 2013, 4-SA 2014, Food Certification International) (https://fisheries.msc.org/en/fisheries/irish-pelagic-sustainability-association-ipsa-western-mackerel/@@view).

Irish Pelagic Sustainability Group [Irish Pelagic Sustainability Group (IPSG) western mackerel pelagic trawl] (1-FA 2009, 1-SA 2010, 2-SA 2011, 3-SA 2012, 4-SA 2013, Food Certification International) (https://fisheries.msc.org/en/fisheries/irish-pelagic-sustainability-group-ipsg-western-mackerel-pelagic-trawl/@@view).

ISF Iceland mackerel [Iceland Sustainable Fisheries] (1-FA 2017, 1-SA 2018, 2-SA 2019, SAI Global) (https://fisheries.msc.org/en/fisheries/isf-iceland-mackerel/@@view).

Mackerel Industry Northern Sustainability Group [MINSA North East Atlantic mackerel] (1-FA 2016, 1-SA 2017, 2-SA 2018, Acoura Marine; 3-SA 2019, Lloyd's Register) (https://fisheries.msc.org/en/fisheries/minsa-north-east-atlantic-mackerel/@@view).

Murmanseld 2 (Russia) [Murmanseld 2 Barents Sea cod and haddock] (1-FA 2020, DNV GL) (https://fisheries.msc.org/en/fisheries/murmanseld-2-barents-sea-cod-and-haddock/@@view).

Northern Ireland Pelagic Sustainability Group [Northern Ireland Pelagic Sustainability Group (NIPSG) Irish Sea-Atlantic mackerel & North Sea herring] (1-FA 2016, 1-SA 2017, Acoura Marine; 2-SA 2019, 3-SA 2019, Lloyd's Register) (https://fisheries.msc.org/en/fisheries/northern-ireland-pelagic-sustainability-group-nipsg-irish-sea-atlantic-mackerel-north-sea-herring/@@view).

Norway (Norw. Seafood Council) [Norway North East Arctic cod] (1-FA 2010, 1-SA 2011, 2-SA 2012, 3-SA 2013, 4-SA 2014, 1-RA 2015, DNV GL) (https://fisheries.msc.org/en/fisheries/norway-north-east-arctic-cod/@@assessments).

Norway lumpfish [NFA Norway ling & tusk and NFA Norway lumpfish] (1-FA 2017, Acoura Marine; 1-SA 2019, 2-SA 2020, DNV GL) (https://fisheries.msc.org/en/fisheries/nfa-norway-ling-tusk-and-nfa-norway-lumpfish/@@assessments).

Norwegian Fishermen's Sales Organization for Pelagic Fish [North East Atlantic mackerel pelagic trawl, purse-seine and handline] (1-FA 2009, 1-SA 2010, 2-SA 2011, 3-SA 2012, 4-SA 2013, DNV GL) (https://fisheries.msc.org/en/fisheries/north-east-atlantic-mackerel-pelagic-trawl-purse-seine-and-handline/@@view).

Ocean Trawlers (Russia) [Barents Sea cod, haddock and saithe] (1-FA 2010, 1-SA 2011, 2-SA 2012, 3-SA 2013, 4-SA 2015, Food Certification International; 1-RA 2016, Acoura Marine) (https://fisheries.msc.org/en/fisheries/barents-sea-cod-haddock-and-saithe/@@view).

Oceanprom (Russia) [Oceanprom Barents Sea cod and haddock] (1-FA 2019, DNV GL) (https://fisheries.msc.org/en/fisheries/oceanprom-barents-sea-cod-and-haddock/@@assessments).

Pelagic Freezer Trawler Association (Netherlands) [Pelagic Freezer Trawler Association North East Atlantic mackerel pelagic trawl] (1-FA 2009, 1-SA 2010, 2-SA 2011, 3-SA 2012, 4-SA 2013, Moody Marine) (https://fisheries.msc.org/en/fisheries/pelagic-freezer-trawler-association-north-east-atlantic-mackerel-pelagic-trawl/@@view).

Ross Sea toothfish [Ross Sea toothfish longline] (1-FA 2010, 1-SA 2011, 2-SA 2012, 3-SA 2013, 4-SA 2014, Moody Marine; 1-RA 2015, Acoura Marine) (https://fisheries.msc.org/en/fisheries/ross-sea-toothfish-longline/@@view).

Saint Malo/Euronor (France) [Compagnie des Pêches Saint Malo and Euronor cod and haddock] (1-FA 2012, 1-SA 2013, 2-SA 2014, MacAlister Elliott and Partners Ltd.; 3-SA 2015, 4-SA 2016, 1-RA 2017, ME Certification Ltd.) (https://fisheries.msc.org/en/fisheries/compagnie-des-peches-saint-malo-and-euronor-cod-and-haddock/@@view).

SARPC toothfish [SARPC toothfish (Réunion)] (1-FA 2013, 1-SA 2014, MacAlister Elliot and Partners; 2-SA 2015, 3-SA 2017, 4-SA 2018, ME Certification; 1-RA 2018, Control Union Pesca) (https://fisheries.msc.org/en/fisheries/sarpc-toothfish/@@view).

Scottish Pelagic Sustainability Group [SPSG Ltd western component of north east Atlantic mackerel] (1-FA 2009, 1-SA 2010, 2-SA 2011, 3-SA 2012, 4-SA 2012, Food Certification International) (https://fisheries.msc.org/en/fisheries/spsg-ltd-western-component-of-north-east-atlantic-mackerel/@@view).

South Georgia Patagonian toothfish [South Georgia Patagonian toothfish longline (South Georgia and the South Sandwich Islands/UK)] (1-FA 2004, 1-SA 2005, 2-SA 2006, 3-SA 2007, 4-SA 2008, 1-RA 2009, Moody Marine) (https://fisheries.msc.org/en/fisheries/south-georgia-patagonian-toothfish-longline/@@view).

Strelets/Eridan (Russia) [Russian Federation Barents Sea cod, haddock and saithe] (1-FA 2014, 1-SA 2015, 2-SA 2016, 3-SA 2017, 4-SA 2018, 1-RA 2019, DNV GL) (https://fisheries.msc.org/en/fisheries/russian-federation-barents-sea-cod-haddock-and-saithe/@@view).

Swedish Pelagic Federation Producer Organisation [SPFPO North East Atlantic mackerel] (1-FA 2011, 1-SA 2012, DNV GL) (https://fisheries.msc.org/en/fisheries/spfpo-north-east-atlantic-mackerel/@@view).

UK Fisheries/DFFU/Doggerbank [UK Fisheries Ltd./DFFU/Doggerbank Northeast Arctic cod, haddock and saithe] (1-FA 2012, 1-SA 2013, 2-SA 2014, MacAlister Elliott and Partners Ltd.; 3-SA 2015, 4-SA 2016, 1-RA 2017, ME Certification Ltd.) (https://fisheries.msc.org/en/fisheries/uk-fisheries-ltd-dffu-doggerbank-northeast-arctic-cod-haddock-and-saithe/@@view).

CHAPTER 6

Conclusions

Abstract Private certification has become a 'separate lane' to implementation of the precautionary approach to fisheries management—it has found its niche as a supplement to implementation by states, assuming a subsidiary but significant role. States work individually and in partnership, primarily in regional fisheries management organizations (RFMOs), to further the implementation of the precautionary approach, but they are in effect assisted by private certification. The visible and non-visible interaction between state action and private certification creates a constructive dynamic that takes the implementation of the precautionary approach further. The iterative and evolving nature of the MSC Certification Scheme ensures further robustness. The Scheme is a 'living' and dynamic implementation mechanism—internally, as an iterative and evolving scheme, and externally, working in tandem with state implementation.

Keywords Ecolabelling · Fisheries certification · Marine Stewardship Council

The overarching research question of this book has been to what extent and how private standards like the MSC Certification Scheme contribute

© The Author(s), under exclusive license to Springer Nature Switzerland AG 2021

G. Hønneland, *Blue Governance in the Arctic and Antarctic*, Palgrave Studies in Maritime Politics and Security, https://doi.org/10.1007/978-3-030-72585-3_6

to the implementation of the precautionary approach to fisheries management. Specifically, I have asked to what extent and in what form the precautionary approach has been incorporated in the MSC Certification Scheme itself, and to what extent and in what way MSC certification has affected target (fisher) behaviour and state practice in the application of the precautionary approach. In this chapter, I summarize my main empirical findings and elaborate on the wider analytical implications of the study. Of special interest is how private certification can, alongside state implementation, contribute as a subsidiary implementation mechanism to influence fisher behaviour and state practice.

Operationalizing and Implementing the Precautionary Approach

In the academic literature, the precautionary approach is split into 'strong' or 'narrow' and 'weak' or 'broad' versions of the principle.[1] The 'strong/narrow' version is normally seen in relation to the reversal of the burden of proof, whereas the weakest/broadest versions prescribe due diligence without defining any quantitative benchmarks for action. In legally binding instruments (the FSA) and non-legally binding ones (the CCRF), as well as in the more practice-oriented literature on the precautionary approach in fisheries management, two versions of the approach emerge: one 'semi-strong and narrow' and one 'semi-weak and broad'.[2] The 'semi-strong/narrow' approach commands reference points and HCRs, whereas the 'semi-weak/broad' version merely envisages typical components of a sound management system, such as stock assessment, access control, technical regulations, enforcement and transparency in decision-making.

The MSC Certification Scheme includes both substantive and procedural requirements to precaution. First, application of the approach is evaluated as such under P3 in fisheries assessments. Second, it serves as a general interpretation device in the application of the MSC Fisheries Standard. Third, the Standard is designed to assist in the operationalization of precautionary reference points. Fourth, certification is not only

[1] See, e.g., Bodansky (2004), Ebben (2011), Gillespie (2007), and Sands and Peel (2018). For a further discussion, see Chapter 2.

[2] See, e.g., Hewison (1996) and Garcia (1994). For a further discussion, see Chapter 2.

a question of the status of the (target) fish stocks. The scheme involves thorough assessment of the fishery's impacts on the marine ecosystem, including bycatch species, ETP species and habitats—as well as management components such as a well-functioning legislative basis and institutional set-up, transparency and stakeholder participation, compliance and enforcement, and review of the management system.

MSC certification has had several precautionary effects in the polar fisheries examined here. It has (i) been an attestation to the appropriateness and soundness of the established reference points and HCRs in the Barents Sea fisheries; (ii) helped to ascertain the appropriateness and soundness of HCRs in the Southern Ocean fisheries; and (iii) been the direct cause of the development and establishment of reference points and HCRs in the Arctic lumpfish fisheries. Improvements have been made to management systems beyond the introduction of reference points and HCRs: new research and information gathering have been conducted in the Southern Ocean fisheries, and efforts have been made to reduce impacts on bottom habitats in the Barents Sea fisheries and on seabirds and marine mammals in the Arctic lumpfish fisheries. In the Northeast Atlantic mackerel fisheries, where the effects of MSC certification are generally weakest, industry groups in the various European countries have joined forces to persuade their respective governments to set TACs in line with scientific recommendations (albeit to no avail so far).

The greatest effect of the MSC Certification Scheme is arguably its extensive *operationalization* of the precautionary approach. First, it offers a global benchmark for establishing reference points and HCRs where previously there was only piecemeal operationalization at the national and RFMO levels. Second, and perhaps its most innovative contribution, the MSC Fisheries Standard operationalizes the precautionary approach not only for the management of target stocks, which has traditionally been the priority of RFMOs and national governments but also for the protection of the marine environment against the impact of fisheries and for critical components and features of the management system.

Private Certification as a Subsidiary Implementation Mechanism

Whereas the narrow version of implementation sees the concept as referring to the transformation or incorporation of international obligations (relating to legally binding rules) and commitments (relating to non-legally binding rules) into national law only (*output*), the broader version also includes changes in other forms of state practice (beyond the introduction and/or amendment of national legislation, including the application of general concepts in specific circumstances, also *output*) and target behaviour aimed at effectuating the requirements of the international instrument/national legislation (*outcome*).[3] 'State practice' is the behaviour or conduct of states, including administrative regulations at the domestic level (insofar as they relate to matters of international law) and how they are interpreted and applied in practice.

In the case studies presented here, I observed the effects of MSC certification in all four clusters of polar fisheries. *Notably, the effects on fisher behaviour were greater than the effects on state practice.* Not unexpectedly, the effects were more significant in fisheries that had had obvious flaws prior to MSC assessment than in those that had not. The Southern Ocean fisheries had suffered from a lack of data and information on fish stocks and the ecosystem—MSC certification led to new research initiatives and voluntary collection of data by fishers, and new epistemic communities were formed. However, I have not been able to document any specific regulations issued by either CCAMLR or national governments that followed directly from MSC certification—the effects on state practice, if indeed any, were insignificant in this fishery. In the Barents Sea fisheries, MSC certification has directed attention to an area that has been neglected by the Joint Norwegian–Russian Fisheries Commission: the fishery's effects on the marine environment, especially bottom habitats. Fishery clients in Russia have joined forces to develop new fishing gear and voluntarily avoid potential new fishing areas that have not yet been properly researched. This is, however, not reflected in, let alone the result of, national or international regulations, and there is no public mention of MSC by either the Joint Commission or national management authorities—the effects on state practice were nil. The situation is similar in the Northeast Atlantic mackerel fisheries, where nothing came out of the 'all-European' fishing industry's efforts to encourage states to agree on sustainable TACs. In the Arctic lumpfish fisheries, on the other hand, states have produced new reference points and HCRs, and at least some regulations can be attributable to MSC certification—the effects on state practice were significant here.

[3] See, e.g., Andresen et al. (2012), List and Rittberger (1998), Victor et al. (1998), and Weiss and Jacobson (1998). For a further discussion, see Chapter 2.

The big picture, however, is that *MSC certification leads to behavioural change among fishers, but not as a result of changed state practice.* Certification becomes a 'short cut' from international obligations and commitments to target behaviour (*outcome*), avoiding the connecting link of national legislation and other forms of state practice (*output*). States are effectively sidelined, as fishers 'take orders' from the MSC (or CABs) directly. And where changes in national regulations *are* required by MSC improvement conditions, fishers find themselves lobbying for stricter regulations, with governments often resisting because they 'do not take orders from the MSC'. Traditional fisheries management is turned on its head: the fishing industry has taken the lead—at least in some instances.

If we look beyond this case study, state-driven implementation has, of course, not become redundant. By and large, fisher behaviour in the Southern Ocean, the Barents Sea, the Northeast Atlantic and in waters under Greenlandic, Icelandic and Norwegian jurisdiction is regulated—and driven—by national legislation, part of which is the result of the implementation of the states' international obligations and commitments. Private certification has become a 'separate lane' to implementation, so to speak—it has found its niche as a supplement to implementation by states, assuming a subsidiary but significant role. States work individually and in partnership, primarily in RFMOs, to further the implementation of the precautionary approach, but they are in effect assisted by private certification. Sometimes certification strengthens measures initiated by states (like client support to the synoptic krill survey in the Southern Ocean); sometimes it supplements those measures (like the voluntary collection of information in all the fisheries studied here). Sometimes certification leads to results that were not even considered (at least openly) by management authorities as a regulatory option (like the development of lighter fishing gear by the Russian MSC clients in the Barents Sea). And sometimes certification does the 'dirty job' and acts when states are oscillating (like fixing reference points and HCRs in the lumpfish fisheries once and for all).

This visible and non-visible interaction between state action and private certification creates a constructive dynamic that takes the operationalization and implementation of the precautionary approach further, step by step. The iterative and evolving nature of the MSC Certification Scheme ensures further robustness. Fisheries are subject to initial certification followed by annual surveillance audits and re-certification every five years, as well as expedited audits whenever developments in the fishery demand them. The MSC Fisheries Standard is revised every five years as well, and with every iteration the bar is raised. The MSC Certification Scheme is a 'living' and dynamic implementation mechanism—internally, as an iterative and evolving scheme, and externally, working in tandem (albeit in parallel more than jointly) with state implementation. Private certification has become a forceful and flexible management device, an organic machinery for change in the world's fisheries.

The handling of the Northeast Atlantic mackerel fisheries was not the MSC's finest hour. It could have become one of its big success stories if the engagement of all the clients with their national authorities had produced the intended effects. As that failed to happen, what remains is the image of a scheme owner that may have gone a bit too far in seeking to keep one of the biggest, classic European fisheries within its portfolio. Here, the MSC itself became more involved than in the other Arctic fisheries assessments, and that it usually gets more generally. Normally, the MSC limits its role to producing the certification requirements, providing technical oversight comments to assessment reports, and issuing certificates. In the Northeast Atlantic mackerel fisheries, however, the MSC involved itself in the harmonization of scores between fisheries and, more importantly, in the prolongation of certificates beyond what is usual. After four years of suspension, the mackerel certificates were reinstated in 2016, despite no progress in the coastal states negotiations, where the stalemate was the reason for suspension in the first place. The deadline for having a coastal states agreement in place was effectively postponed until 2021, ten years after the Independent Adjudicator had ruled against certification of the Northeast Atlantic mackerel fishery due to the lack of a coastal states agreement. Meanwhile, the fisheries remained certified—until the stock itself fell below acceptable levels and the fishery was suspended for that reason in 2019.

State Practice Revisited

I deliberately chose not to investigate the effects of private certification on the legal *status* of the precautionary approach, where the big question is whether the approach has achieved status as customary international law. The empirical scope is far too narrow to enable me to make any pronouncements on that. Even if my conclusion had been that MSC certification has extensively influenced state practice in the fisheries examined here (which it has not), that too would have been a completely insignificant contribution to the assessment of the legal status of the precautionary approach under international law. For a rule (or a principle) to achieve status as customary international law, it must reflect a practice that is widespread and representative ('general'), consistent and accompanied by a recognition that the practice is based on a legal obligation, rather than mere usage or custom. A first step would be to examine how widespread the practice is.

By mapping developments in selected polar fisheries, I have made a small contribution to our understanding of the extent to which the precautionary approach has been applied in fisheries around the world, and to our understanding of the diffusion of the approach in state practice. I have already concluded that the effects of MSC certification on state practice—measured in terms of changes in fisheries management as a result of improvement conditions attached to fishery certificates—have been modest, at best. However, that does not exclude the possibility that certification may have had wider effects. When a fishery is certified *without* a condition on a specific parameter, e.g. because an explicit HCR is already in place, that implies that the relevant criteria were met prior to MSC assessment. The effect of certification is then that the negative consequences of abandoning such an HCR become greater. Whereas previously such an act might not have much consequence beyond criticism from environmental NGOs, it would now lead to fisheries losing their MSC certificate and access to important markets. Hence, MSC certification may have greater effects on state practice than just leading to improvements in fisheries with identifiable flaws: it also cements best practice.

In order to improve our understanding of the contribution of private certification to state practice, studies must be conducted with a far wider empirical scope. More fisheries from different parts of the world must be included, as well as other mechanisms than the fulfilment of conditions, such as the preserving effect of certification on best practices. The lines of causation are intricate, and a particularly interesting topic for further research would be to what extent extensive changes in fisher behaviour following directly from certification (without prior changes in national or international regulations) may gradually have effects on state practice as well. Will states follow suit with appropriate regulations when new and more sustainable fishing practices establish themselves as a result of private certification? If so, we may see closer coordination of private and public initiatives. That would make a discussion of private certification and state practice more pertinent as well.

References

Andresen, Steinar, Elin Lerum Boasson and Geir Hønneland, 'An International Environmental Policy Takes Shape', in Steinar Andresen, Elin Lerum Boasson and Geir Hønneland (eds), *International Environmental Agreements: An Introduction*, Abingdon: Routledge, 2012.

Bodansky, Dan, 'Deconstructing the Precautionary Principle', in David D. Caron and Harry N. Scheiber (eds), *Bringing New Law to Ocean Waters*, Leiden: Martinus Nijhoff, 2004.

Ebben, Thomas, 'The Implementation of the Precautionary Approach in International Fishery Law: A Move towards "Green" Fishery', 15 *New Zealand Journal of Environmental Law* 2011, pp. 113–146.

Garcia, Serge Michel, 'The Precautionary Principle: Its Implications in Capture Fisheries Management', 22 *Ocean & Coastal Management* 1994, pp. 99–125.

Gillespie, Alexander, 'The Precautionary Principle in the 21st Century: A Case Study of Noise Pollution', 22 *International Journal of Marine and Coastal Law* 2007, pp. 61–87.

Hewison, Grant J., 'The Precautionary Approach to Fisheries Management: An Environmental Perspective', 11 *International Journal of Marine & Coastal Law* 1996, pp. 301–332.

List, Martin and Volker Rittberger, 'The Role of Intergovernmental Organizations in the Formation and Evolution of International Environmental Regimes', in Arild Underdal (ed.), *The Politics of International Environmental Management*, Dordrecht: Kluwer Academic, 1998.

Sands, Philippe and Jaqueline Peel with Adriana Febra and Ruth MacKenzie, *Principles of International Environmental Law*, fourth edition, Cambridge: Cambridge University Press, 2018.

Victor, David G., Kal Raustiala and Eugene B. Skolnikoff (eds), *The Implementation and Effectiveness of International Environmental Commitments: Theory and Practice*, Cambridge, MA: MIT Press, 1998.

Weiss, Edith Brown and Harold K. Jacobson (eds), *Engaging Countries: Strengthening Compliance with International Environmental Accords*, Cambridge, MA: MIT Press, 1998.

FSC
www.fsc.org
MIX
Papier | Fördert gute Waldnutzung
FSC® C083411

Kolibri 360 GmbH
Ferdinand-Jühlke-Straße 7
99095 Erfurt, Deutschland
produktsicherheit@kolibri360.de